Advanced Theories and Computational Approaches to the Electronic Structure of Molecules

NATO ASI Series

Advanced Science Institutes Series

A series presenting the results of activities sponsored by the NATO Science Committee, which aims at the dissemination of advanced scientific and technological knowledge, with a view to strengthening links between scientific communities.

The series is published by an international board of publishers in conjunction with the NATO Scientific Affairs Division

A	Life Sciences	Plenum Publishing Corporation
B	Physics	London and New York
C	Mathematical and Physical Sciences	D. Reidel Publishing Company Dordrecht, Boston and Lancaster
D	Behavioural and Social Sciences	Martinus Nijhoff Publishers
E	Engineering and Materials Sciences	The Hague, Boston and Lancaster
F	Computer and Systems Sciences	Springer-Verlag
G	Ecological Sciences	Berlin, Heidelberg, New York and Tokyo

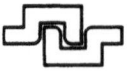

Series C: Mathematical and Physical Sciences Vol. 133

Advanced Theories and Computational Approaches to the Electronic Structure of Molecules

edited by

Clifford E. Dykstra

School of Chemical Sciences,
University of Illinois, Urbana, Illinois, U.S.A.

D. Reidel Publishing Company

Dordrecht / Boston / Lancaster

Published in cooperation with NATO Scientific Affairs Division

Proceedings of the NATO Advanced Research Workshop on
Vectorization of Advanced Methods for Molecular Electronic Structure
Colorado Springs, Colorado, U.S.A.
September 25-29, 1983

Library of Congress Cataloging in Publication Data

NATO Advanced Research Workshop on Vectorization of Advanced Methods for Molecular
Electronic Structure (1983: Colorado Springs, Colo.)
Advanced theories and computational approaches to the electronic structure of molecules.

(NATO ASI series. Series C, Mathematical and physical sciences; vol. 133)
"Proceedings of the NATO Advanced Research Workshop on Vectorization of Advanced
Methods for Molecular Electronic Structure, Colorado Springs, Colorado, U.S.A., September
25-29, 1983"—T.p. verso.
"Published in cooperation with NATO Scientific Affairs Division."
Bibliography: p.
Includes index.
1. Molecular structure—Congresses. 2. Electronic structure—Congresses. I. Dykstra,
Clifford E. II. North Atlantic Treaty Organization. Scientific Affairs Division. III. Title.
IV. Series: NATO ASI series. Series C, Mathematical and physical sciences; vol. 133.
QD461.N34   1983   541.2'2   84-15918
ISBN-13: 978-94-009-6453-2   e-ISBN-13: 978-94-009-6451-8
DOI: 10.1007/978-94-009-6451-8

Published by D. Reidel Publishing Company
P.O. Box 17, 3300 AA Dordrecht, Holland

Sold and distributed in the U.S.A. and Canada
by Kluwer Academic Publishers,
190 Old Derby Street, Hingham, MA 02043, U.S.A.

In all other countries, sold and distributed
by Kluwer Academic Publishers Group,
P.O. Box 322, 3300 AH Dordrecht, Holland

D. Reidel Publishing Company is a member of the Kluwer Academic Publishers Group

CONTENTS

PREFACE

That there have been remarkable advances in the field of molecular electronic structure during the last decade is clear not only to those working in the field but also to anyone else who has used quantum chemical results to guide their own investigations. The progress in calculating the electronic structures of molecules has occurred through the truly ingenious theoretical and methodological developments that have made computationally tractable the underlying physics of electron distributions around a collection of nuclei. At the same time there has been considerable benefit from the great advances in computer technology. The growing sophistication, declining costs and increasing accessibility of computers have let theorists apply their methods to problems in virtually all areas of molecular science. Consequently, each year witnesses calculations on larger molecules than in the year before and calculations with greater accuracy and more complete information on molecular properties.

We can surely anticipate continued methodological developments of real consequence, and we can also see that the advance in computational capability is not about to slow down. The recent introduction of array processors, multiple processors and vector machines has yielded a tremendous acceleration of many types of computation, including operations typically performed in quantum chemical studies. Utilizing such new computing power to the utmost has required some new ideas and some reformulations of existing methods. For this reason, a NATO Advanced Research Workshop was organized to look at the experiences of users of these new computers in relation to electronic structure methods, particularly electron correlation treatments which are invariably the most computationally demanding.

The workshop discussion sessions were organized around specific topics with the idea of assessing state-of-the-art approaches and identifying common or unifying features suited to the new generation of computers. For the most part, the papers included in this volume of proceedings follow the workshop's organization and incorporate ideas and conclusions brought out in open discussion.

The workshop, officially titled "Vectorization of Advanced Methods for Molecular Electronic Structure. An Assessment of State-of-the-Art Electron Correlation Theories and Unifying Features Suited to Vectorized Computation", was held under the auspices of the NATO Scientific Affairs Division. We gratefully acknowledge their generous support and their interest in this meeting. We should also like to thank Floating Point Systems, Inc. for their special grant of funds in support of advanced computational methods in quantum chemistry.

The workshop was held during the week of September 25, 1983 in Colorado Springs, Colorado. The organizing committee consisted of Professor Reinhart Ahlrichs, Professor Wilfried Meyer and myself. The School of Chemical Sciences at the University of Illinois provided certain resources for the organization and planning of this workshop. The Colorado Springs Convention Bureau and the conference staff of the Antlers Plaza Hotel provided invaluable assistance for the arrangements at the meeting site.

January, 1984   Clifford E. Dykstra
                                             Urbana, Illinois

CHEMICAL COMPUTATIONS ON AN ATTACHED PROCESSOR: QUANTUM CHEMISTRY APPLICATIONS

Thom. H. Dunning, Jr. and Raymond A. Bair

Theoretical Chemistry Group, Chemistry Division,
Argonne National Laboratory, Argonne, Illinois 60439

The Floating Point Systems, Inc. Model 164 (FPS-164) attached processor is a high speed, pipelined, parallel processor designed for large-scale scientific computations. Benchmark studies indicate that the FPS-164 is more than an order of magnitude faster than a DEC VAX 11/780 and comparable in speed to an IBM 3033 or CDC Cyber 175 mainframe computer. Vectorization and optimization techniques applied to electronic structure codes for the FPS-164 can lead to further improvements, e.g, GVB164, a generalized valence bond program, runs 30 to 40 times faster on the FPS-164 than on the VAX 11/780.

I. INTRODUCTION

In the 1960s and early 1970s most theoretical chemists used the large mainframe computers available in their institution's central computing facility for all calculations. While this offered the opportunity of carrying out large-scale calculations, in fact, the large operating cost associated with the general purpose computing facility discouraged most large-scale applications unless they were heavily subsidized. In 1972, Schaefer & Miller implemented a number of codes for chemistry computations on a Harris/4 minicomputer from Harris Corporation. They found that such a special purpose computer was a cost-effective alternative to the large mainframe computers in the central facility (1).

Use of minicomputers in large-scale chemical computations was given further impetus in 1976 with the introduction of the VAX 11/780 super-minicomputer by Digital Equipment Corporation.

1

C. E. Dykstra (ed.),
Advanced Theories and Computational Approaches to the Electronic Structure of Molecules, 1–12.
© 1984 by D. Reidel Publishing Company.

The VAX 11/780 was not only substantially faster than the
Harris/4 but in addition many of the architectural limitations
that restricted the applications of the latter machine were
eliminated in the VAX 11/780. Since 1976 use of VAX 11/780
super-minicomputers or its equivalents (the Harris 800 from
Harris Corporation, the PE 3240 and 3250 from Perkin-Elmer
Corporation, and the 4331 and 4341 from IBM Corporation) in
chemical computations has proliferated; today most major theoretical chemistry groups as well as many chemistry departments
possess super-minicomputers for scientific computations, data
analysis, etc.

Today, however, even the VAX 11/780 super-minicomputer is
inadequate for more than routine applications. A typical
large-scale calculation at the forefront of research in theoretical chemistry may require several days to complete on such
a machine – a research program simply cannot be effectively
pursued in this manner. Within the last year a new generation
of super-minicomputers has been introduced – by Data General
(the MV 10000), Gould/SEL (the Concept 32/87), Prime (Prime
9950), Harris (Harris 1000), IBM (IBM 4361 and 4381) and Perkin
Elmer (3250XP); see reference 2. These machines all promise
speeds several times that of the VAX 11/780 at comparable
costs. While the introduction of these new computers will help
alleviate the shortfall of computing power in theoretical chemistry research, it should be noted that they are still several
times slower than common mainframe computers, such as the IBM
3033 or the CDC Cyber 175, and as much as two orders of magnitude slower than the supercomputers in widespread use, the
Cray-1S and the CDC Cyber 205.

In mid-1982 the Theoretical Chemistry Group at Argonne
National Laboratory acquired a Floating Point Systems, Inc.
Model 164 (FPS-164) Attached Scientific Processor. The FPS-164
is a high speed, vector (parallel, pipelined) processor
designed for large-scale scientific computations – in effect, a
mini-supercomputer. For large-scale scientific computations
the FPS-164 provides performance comparable to that of modern
mainframe computers at a cost comparable to that of a super-
minicomputer system. However, like all vector processors,
e.g., the Cray-1S and CDC Cyber 205 supercomputers, to achieve
this performance care must be exercised in developing and
implementing codes on the FPS-164.

The FPS-164 has been discussed in some detail in earlier
papers by Bair & coworkers (3). In this paper we briefly review the hardware and software features of the FPS-164, present
the results of selected benchmark calculations including that
of a production benchmark, and briefly discuss the general
principles which are important in coding applications for the

FPS-164. The last topic will be discussed in greater detail in a separate publication (4).

II. THE FLOATING POINT SYSTEMS, INC. MODEL 164 ATTACHED SCIENTIFIC PROCESSOR

The major hardware and software features of the Floating Point Systems, Inc. Model 164 are described briefly below.

Hardware Specifications

Precision: The central processor uses a full 64 bits in all floating point computations, corresponding to an accuracy of 15 decimal digits. While lesser precision can be used in selected applications in quantum chemistry, many applications require the high precision offered by full 64 bit arithmetic.

Speed: The cycle time of the FPS-164 is 182 nsec. Since a floating point addition and a floating point multiplication can be carried out each cycle, this gives a peak rate of 11 megaFLOPs. This can be compared to megaFLOPS rates of approximately 0.5 for the VAX 11/780, and 150 for a Cray-1S. Based solely on the megaFLOPS rate we would thus expect that the FPS-164 would be as much as 20 times faster than a VAX 11/780 and less than 15 times slower than a Cray-1S for numerically intensive applications.

Memory: The FPS-164 presently supports 7.25 Mwords of main memory and 32 kwords of random access table memory. This is far in excess of that available on super-minicomputers and exceeds that of all but the latest supercomputers. Unfortunately, the present high cost of main memory on the FPS-164 ($176,000/Mword) discourages its use. The address space of the FPS-164 is actually capable of supporting 16 Mwords of main memory.

Disk Drives: The FPS-164 supports a maximum of 24 disk drives on six separate controllers. As the present drives each have a capacity of 16 Mwords, the total disk storage provided is nearly 400 Mwords. The transfer rate to the present disks is 110 kwords/sec; this rate is limited by the transfer rates of the disks themselves, as the I/O bus is capable of a transfer rate of 5.5 Mwords/sec. Further, the software interfaces used to drive the disk transfers reduces this rate by at least 15%. Design rules for high performance computers (5) suggest, and our experience indicates, that the disk transfer rates are below those needed to sustain the central processing unit in applications involving heavy I/O to the disk drives. Floating Point Systems, Inc. is said to be addressing this problem.

Host Interface: The FPS-164 is interfaced to the VAX 11/780 host through the Unibus. The maximum transfer rate on the Unibus is 140 kwords/sec; inefficiencies in the VAX-FPS driver software reduce this by more than 40%. The slow transfer rate over the Unibus prevents applications which require substantial data transfer between the host and the FPS-164.

Software Specifications

Compiler: The compiler for the FPS-164 is a standard FORTRAN-77 compiler. At the highest optimization level it is able to take advantage of the special architectural features of the machine through pipeline optimization techniques. The present version of the compiler (E00 Release) is fairly reliable although not all of our application subroutines will compile correctly at the pipeline optimization level.

Mathematical Subroutine Library: The FPS-164 has a standard library of nearly 400 Fortran-callable mathematics subroutines. These routines are written in assembly language. The library includes the primitive BLAS vector operations and a few vector and matrix routines from the LINPACK and EISPACK packages and, thus, provides fast vector and matrix operations without resorting to specialized assembly language coding.

Operating System: The latest release of the operating system for the FPS-164 (SJE, E00 Release) supports permanent files on the attached processor disks and includes a roll-in/ roll-out facility (not yet fully debugged) which allocates time slices between jobs. The SJE operating system provides many features of importance to computational chemists. In addition to the two features mentioned above, the SJE command language supports jobs containing an arbitrary number of program executions as well as access to files on the host computer. For production computations, the FPS-164 can operate like an independent computer. These features separate the FPS-164 from other attached processors which can only operate as "subroutine boxes" under the control of a specially coded host program.

The software for the FPS-164 also includes an interactive debugger, assembler, object linker and object librarian.

III. RESULTS OF BENCHMARKS FOR THE FLOATING POINT SYSTEMS, INC. MODEL 164 ATTACHED PROCESSOR

To assess the speed of the FPS-164 a number of benchmarks were run on the machine. The results of these tests are summarized in Figures 1-3.

In the matrix benchmark · a number of real symmetric matrices, developed from the recommendations of Liu (6) were diagonalized. Using codes written in Fortran the FPS-164 is seen, Figure 1, to be nearly 11 times faster than the ubiquitous VAX 11/780, 7 times faster than the Harris 800, almost as fast as the IBM 3033 and the CDC Cyber 175. If the FORTRAN matrix diagonalization routine is replaced with a call to the identical assembly coded routine in the mathematic subroutine library on the FPS-164, the speed of the FPS-164 increases by a factor of 2.5. This is indicative of the raw computing power of the FPS-164.

The vector benchmark (7) involved the solution of a system of linear equations. Again using codes written entirely in FORTRAN, the FPS-164 is seen to be approximately 1/2 the speed of the CDC 7600, the supercomputer of the early and mid 1970s, 1/5-1/8 of the speed of the supercomputers of the late 1970s and early 1980s, the Cray-1S and Cyber 205, and 1/14 the speed

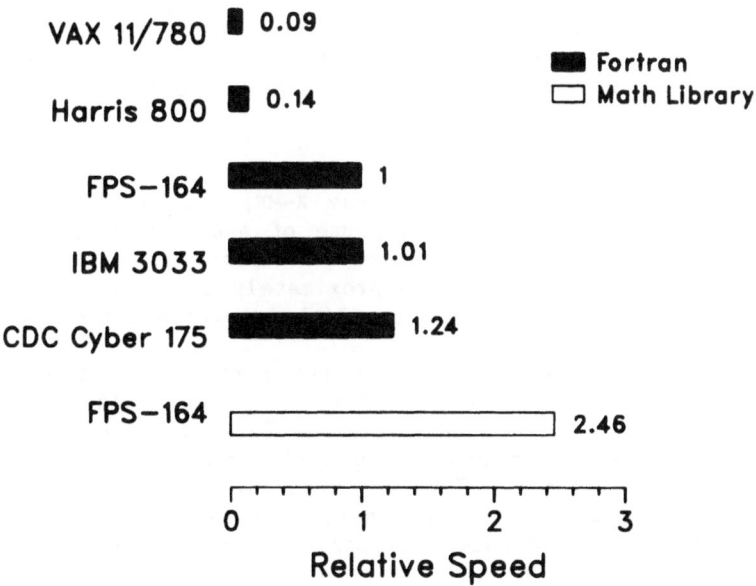

Figure 1. Matrix benchmark: relative times to diagonalize ten real symmetric matrices of rank 10, 20, 30, ..., 100 using the EISPACK RS routine for the FPS-164 (assigned a unit time) and a number of widely used super-minicomputers and mainframe computers. Each matrix contributes to the total CPU time as $(100/N)^3$.

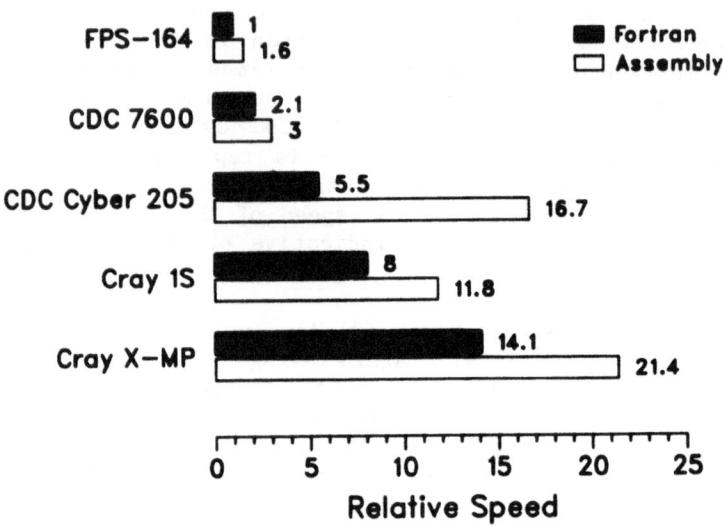

Figure 2. Vector benchmark: relative times to solve a single 100x100 system of linear equations using the LINPACK and BLAS routines for the FPS-164 (assigned a unit time) and a number of currently available super-computers (7).

of the recently delivered Cray X-MP; see Figure 2. With the exception of the Cyber 205, use of assembly coded versions of the same subroutine improves the performance of all of the machines by a factor of approximately 1.5 (the performance of the Cyber 205 is improved by slightly over a factor of 3).

Of course, in evaluating the performance of any new computer system the most important measurement is the amount of production work that can be done. To assess the performance of the FPS-164 in our production environment an additional benchmark was developed. This benchmark involved the calculation of the energy for the vinyl radical (C_2H_3) – calculation of the integrals for a double zeta polarized basis set (43 functions), computation of a five-pair generalized valence bond wavefunction (32 configurations), transformation of the integrals from the atomic orbital basis to the molecular orbital basis, and, finally, calculation of a CI wavefunction consisting of all single and double excitations relative to the first natural configuration (23,550 configurations). Such calculations are an essential part of the theoretical chemistry research program at ANL. The results of the production benchmark are summarized in Figure 3. Overall, the benchmark ran nearly 11 times faster

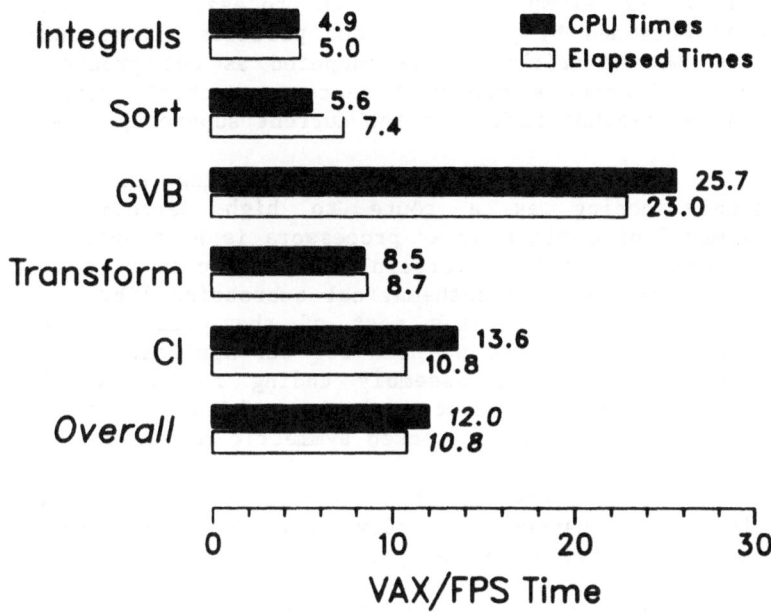

Figure 3. Production benchmark: comparison of FPS-164 and VAX 11/780 times for the production calculation on the vinyl radical (October 28, 1983).

on the FPS-164 than on the VAX 11/780. Considering only the central processor times, the FPS-164 is a full 12 times faster than the VAX 11/780. Thus, the production benchmark also puts the FPS-164 on a par with the IBM 3033 and CDC Cyber 175 mainframe computers. Further, continuing work on the quantum chemistry codes is expected to increase the speed of the production calculation by at least 40-50%.

IV. OPTIMIZATION OF CODES FOR THE FLOATING POINT SYSTEMS, INC. MODEL 164 ATTACHED PROCESSOR

Since we expect to port our electronic structure codes to higher performance supercomputers, such as the Cray-1 and Cyber 205, as they become available, we decided to adopt optimization schemes which would be partially, if not completely, compatible with a generic vector processor. Although such code will not have the best possible performance on any one particular computer, we expect to be able to obtain vector performance on the supercomputers available now and in the near future. Of course, when selecting between equivalent algorithms, we chose those that are best suited to the architecture of the

FPS-164. For example, the maximum processing speed on the FPS-164 is obtained by performing dot products. Consequently much of our matrix arithmetic is computed as dot products, rather than the "vector = scalar × vector + vector" operation which would be somewhat faster on the current supercomputers.

For the above reasons we have also excluded massive assembly coding as a route to high performance. This "hallmark" of earlier array processors is best suited to static applications, which is certainly not the case in quantum chemistry. Besides, the mathematical subroutine library available for the FPS-164 contains most of the primitive vector and matrix operations that we need for our applications. We have, however, resorted to assembly coding for those classes of general matrix and vector operations which were not available in the library (e.g. for packed symmetric matrices).

This philosophy has been developed into a strategy for implementing quantum chemistry codes on the FPS-164. The strategy has five essential points:

(1) Code loops to optimize pipeline performance. This process involves eliminating branches, conditional code, and other pipeline breaking constructs from the innermost loops of a program. The same techniques usually optimize loop performance in supercomputers, although the FPS-164 can pipeline (vectorize) a much wider range of loops than the Cray and CDC Cyber computers.

(2) Replace simple vector operations with calls to the mathematical library routines, preferably the BLAS subroutines. Ideally, a compiler would create optimal code for simple vector operations. Since this is not the case for the FPS-164, and many other computers, this ubiquitous package offers the best alternative to assembly language programming. Compound vector operations are usually done most efficiently in Fortran rather than with several vector subroutine calls.

(3) Perform as much matrix arithmetic as possible with the mathematical library routines, preferably the LINPACK or EISPACK subroutines. Matrix operations (e.g. matrix multiplication, eigenvalue systems, linear equation solutions) often have a high ratio of arithmetic operations to memory accesses, which can be best exploited in these optimized subroutines.

(4) Minimize integer arithmetic, floating point divisions, and conditional code. Table-driven algorithms and pre-computed pointers provide avenues to this kind of optimization which enhances the speed of a program on almost any computer. These optimizations are important with the FPS-164 which has no

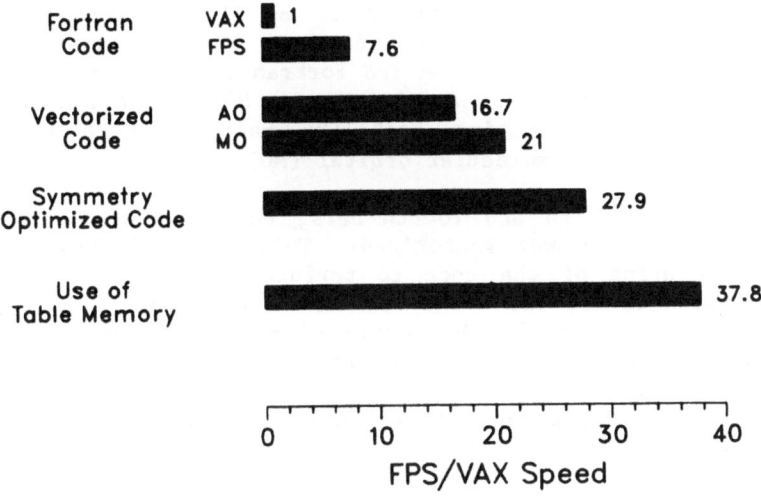

Figure 4. Optimization of the GVB(pp) code for the FPS-164; for an explanation of the various optimization stages, see the text.

dividing hardware (a reciprocal approximation with series solution is used) and which performs integer arithmetic with the floating point unit.

(5) Use asynchronous input/output with large buffers. Unformatted disk I/O is much more efficient with large buffers than with small buffers. Asynchronous I/O allows much of the I/O time to be covered by computations. Spreading sequential files onto several different disks (preferably with independent controllers) reduces the head positioning latency in disk accesses.

As an example of these techniques, Figure 4 shows the improvement obtained at various stages of optimization of the generalized valence bond wavefunction program, GVB164 (8). The benchmark calculations compared in the figure involve the computation of a GVB wavefunction for the vinyl radical using 43 basis functions and 5 correlated electron pairs (as described in reference 3a and the production benchmark). The time ratios are CPU times for the wavefunction optimization section of GVB164. For the first bar, the VAX 11/780 version of the program has been assigned a ratio of 1. The remaining bars display ratios of FPS-164 versions to the original VAX version.

For the second bar, the VAX Fortran code was implemented on the FPS-164 with a minimum of changes. One VAX optimization iteration took 170 seconds while an FPS iteration took only 22.4 seconds, a factor of 7.6 times faster. Ratios of this size are obtained when the FPS Fortran compiler can effectively optimize the program. At this stage, 85% of the iteration time was spent constructing Fock-like matrices over the atomic orbital (AO) and molecular orbital (MO) bases.

In the third and fourth bars, the construction of the AO and MO matrices was vectorized. This involved straightforward restructuring of the code to perform dot products and matrix multiplications with the mathematical library subroutines. Vectorization of the AO matrix element construction and then that for the MO matrix elements increases the performance of the GVB code by factors of 2.2 and 1.3, respectively. At this stage the FPS program ran 21 times faster than the original VAX version.

Up to this point, the density matrices contained a number of zero elements when the orbitals were of different symmetry. By eliminating the zero elements common to all of the density matrices and the corresponding two electron integrals, the number of arithmetic operations was reduced. Vinyl radical has a plane of symmetry, resulting in a reduction of the iteration time from 8.1 to 6.1 seconds. The symmetry adapted version of the program is now 27.9 times faster than the original VAX version. In total, a factor of 3.7 improvement has been obtained by optimizing the original FPS code using entirely portable techniques.

In the last stage, represented by the final bar in Figure 4, the density matrices for the AO matrix element construction were stored in the table memory of the FPS-164. Table memory is a separate memory of 32 kwords which can be accessed simultaneously with main memory. Use of table memory effectively doubles the speed of this process, resulting in an overall improvement of a factor of 1.4. The GVB code now runs 37.8 times faster on the FPS-164 than on the VAX 11/780. Unfortunately, this final optimization gives essentially no improvement in the elapsed time for the calculation, since the AO matrix element construction is limited by the rate at which the two electron integral list can be read from the FPS-164 disks. In terms of elapsed time, the calculation of the GVB wavefunction for the vinyl radical is 23 times faster on the FPS-164 than the corresponding calculation on the VAX (27 times faster for just the wavefunction optimization section). This non-portable optimization technique is a switchable option in the program.

V. SUMMARY

Attached processors, like the FPS-164, are a viable means of upgrading a super-minicomputer system to obtain mainframe performance, while preserving the cost-effectiveness of the minicomputer system. The FPS-164 software (FORTRAN-77 compiler, mathematical subroutine library, file system, etc.) provides a powerful set of tools for implementing electronic structure codes in a portable manner. It is expected that many codes written for the FPS-164 attached scientific processor can be migrated to supercomputers, retaining vector performance. The Theoretical Chemistry Group at Argonne National Laboratory has implemented a complete system of electronic structure codes on the FPS-164, which perform 11-12 times faster than the VAX 11/780 in production benchmark calculations. In particular, the generalized valence bond code, GVB164, was vectorized quite sucessfully, realizing 23 times the VAX speed (elapsed time).

ACKNOWLEDGMENT

We wish to thank our colleagues, Drs. Ron L. Shepard and Russell M. Pitzer, for the use of the integrals, transformation and configuration interaction programs in the production benchmark calculations.

Work performed under the auspices of the Office of Basic Energy Sciences, Division of Chemical Sciences, U. S. Department of Energy, under Contract W-31-109-Eng-38.

REFERENCES

1. Pearson, P. K., Luccese, R. R., Miller, W. H., and Schaefer, III, H. F., 1977, in "Minicomputers and Large Scale Computation", Ed. Lykos, P., ACS Symposium Series 57, Washington, DC, pp. 171-190.
2. Killmon, P., 1983, Computer Design 22, pp. 167-185.
3. (a) Bair, R. A. and Dunning, Jr., T. H., J. Comput. Chem. (in press); (b) Shepard, R., Bair, R. A., Eades, R. A., Wagner, A. F., Davis, M. J., Harding, L. B., and Dunning, Jr., T. H., Intern. J. Quantum Chem. (in press).
4. Bair, R. A. (to be published).
5. Lincoln, N., 1983, IEEE Computer 16, pp. 38-47.
6. Liu, B. in "Numerical Algorithms in Chemistry: Algebraic Methods", NRCC Workshop, Report LBL-8158, Lawrence Berkeley Laboratory, University of California, Berkeley, California, 1978, pp. 49-53.
7. Dongarra, J. J., Moler, C. B., Branch, J. R., and Stewart, G. W., 1979, LINPACK User's Guide, Siam, Philadelphia.

8. GVB164 is the FPS-164 version of GVB2.5, written in 1982
    and 1977, respectively, by R. A. Bair. These programs are
    extensively modified versions of GVBTWO (written by F. W.
    Bobrowicz and W. R. Wadt in 1973).

CONSIDERATIONS IN VECTORIZING THE CI PROCEDURE

Charles W. Bauschlicher, Jr.

NASA Ames Research Center
Moffett Field, California 94035

The ability to square a lower triangular matrix, to fold a square matrix into triangular form, and an efficient matrix multiply are the three important utilities for vectorization of CI. Molecular symmetry treatments must be fully integrated into the program organization.

DISCUSSION

The technique of configuration interaction, CI, is the most widely used and has proven to be the most accurate post-Hartree-Fock approach for the investigation of the properties of small molecules. Throughout the 1960's and early 1970's the conventional CI, one in which the Hamiltonian matrix is formed and stored on disk, was the backbone of CI calculations. Also in use were very powerful special purpose direct CI techniques, ones in which the wavefunction is directly evaluated from the integrals. The conventional CI has the advantage that there are no restrictions on the individual configuration state functions, CSF's, which are used, but it suffers from the severe limitation that all of the non-zero Hamiltonian matrix elements must be stored on disk. The direct CI techniques suffered from the limited form of the wave-functions which could be treated, which was most commonly, single and double excitations away from a single reference CSF. These direct CI techniques where capable of handling classes of problems an order of magnitude larger than the conventional CI techniques.

The late 1970's saw a breakthrough in the direct CI technology (see for example ref. 1-4, and references therein); the form of the wavefunction was expanded to include single and double excita-

C. E. Dykstra (ed.),
Advanced Theories and Computational Approaches to the Electronic Structure of Molecules, 13–18.
© 1984 by D. Reidel Publishing Company.

tions from any arbitrary list of reference CSF's. Extensive work
with conventional CI programs had suggested that this form of the
wavefunction was ideal for solving most classes of problems. This
improvement radically changed the problems which were tractable.
This breakthrough in direct CI technology was further enhanced
by the fact that many components of the direct CI are intrinsically vectorizable and, as such, large speedups are possible on
the current generation of vector machines.

A CI calculation can be divided into four parts: the transformation of integrals, the evaluation of the symbolic formulae,
the iterative determination of the wavefunction and the use of the
wavefunction to determine properties. The evaluation of the symbolic formulae has not received much attention in the vectorizing
process and it does not appear that this process is intrinsically
vectorizable. This combined with the fact that the time needed
for this step is usually small and its cost is amortized over all
the points computed, is probably the reason little attention has
been paid to this aspect of the process and we do not consider it
here. The calculation of properties, either for a single wavefunction or between two wavefunctions, is most directly performed by
first computing a density matrix. For the case of a single wavefunction, this density matrix is formed quite simply from the same
code which is used to compute the one electron contribution to the
wavefunction. For the case of two different wavefunctions, the
calculation of the transition density matrix is very similar to
the single wavefunction case; in both cases the time to form the
density matrix is very similar to the single wavefunction case;
in both cases the time to form the density matrix is very small.
Since the vectorization of this process is very similar to the
evaluation of the CI wavefunction, it is not discussed further.

The transformation of the integrals and the evaluation of
the wavefunction are the time consuming steps in the CI procedure
and both are vectorizable. We consider each one separately. The
four-index out-of-core transformation, as described by Yoshimine (5),
is reduced to a series of two index transformations. If the inte-
grals over the NAO atomic basis functions are denoted by $(pq|rs)$
and the matrix of orbital coefficients by C(NAO,NMO), the procedure is to transform for all possible pq's on the indices rs.
Let the transformed rs be denoted by ij, then $ij = C^\dagger rs\ C$. The
block of integrals is then sorted on the ij index and when the
first half transform is complete, the pq blocks are transformed
for each ij; $kl = C^\dagger pq\ C$, yielding the $(ij|kl)$ integrals over the
molecular orbitals. Clearly, the two index transformation
ij(NMO,NMO) = $C^\dagger$(NMO,NAO) rs(NAO,NAO) C(NAO,NMO), is two matrix
multiplies, MXM. On the CRAY computers, the MXM is the fastest
operation and very high rates of performance should be expected
for the two index transformations.

This is not as straight forward as the formal discussion might imply. In order to minimize the disk storage, only the unique integrals are stored; the rs's are not stored as square, but rather only the lower triangle, $r \geq s$. On a scalar machine, the two index transformation is coded to use only the lower triangle of atomic integrals. One could consider vectorizing such an algorithm or to first square the integrals and perform MXM's. From FORTRAN neither approach will work well, and to fully utilize the CRAY both methods would have to be coded in CAL. (Note that a FORTRAN MXM runs at about 40MFLOP's, while a CAL coded routine performs at about 150MFLOP's.) Since there is an efficient system supplied MXM and many cases of naturally square integral blocks arise in a symmetrized transformation and in the evaluation of the wavefunction, the squaring of the integrals offers a simple and more general approach. The MXM is of the order n^3 and the squaring is of order n^2, thus it would appear that squaring of the integrals should require an insignificant amount of time. However, the MXM runs at ~150MFLOP's and scalar code can run at ~5MFLOP's, thus even for a reasonably sized basis set, the squaring could require about as much time as the MXM. Clearly, some thought must be given to how to square the integrals. Since one is repeatedly squaring the same size matrix, it is possible to build two scatter pointer arrays, one which describes the scattering of the original integrals into the lower triangle and one into the upper triangle. For small matrices this seems to work quite well and will work even better as hardware scatter/gather become available on the CRAY computers. For larger matrices, one row of the original integrals can be efficiently moved into a row and into a column of the square integral matrix. (The short initial rows represent only a small portion of the work and, as such, the poor performance of those does not effect the overall efficiency). Both supply a method to effectively eliminate the time of the n^2 squaring process. Thus, one can achieve the expected high performance rates.

In the above discussion symmetry has been neglected; however, by considering the simple example of two symmetries with an equal number of basis functions, n, the dramatic effect of symmetry can be observed. If symmetry is neglected, there are $(2n*(2n+1)/2)$ two index transformations in the first half transform and an equal number in the second half. Each transformation is of order $(2n)^3$ (neglecting the fact that the two index transformation is in fact two MXM's), or the total amount of work is $32n^5 + 16n^4$. This is compared to the case the integrals have been computed over symmetry orbitals; the non-zero blocks of integrals are $(aa|aa)$, $(bb|bb)$, $(ab|ab)$ and $(aa|bb)$, yielding a total transformation of the order of $4n^5 + 2n^4$ which is an eightfold reduction in work. The dimension of each MXM is smaller in the case with symmetry, but on a CRAY, the MXM rapidly reaches the very high performance rates, such that the reduced sice of the individual MXM's represents no additional problems. Clearly, the inclusion of

symmetry represents a major speedup in the transformation at the
cost of a more sophisticated sort in which the symmetry blocking
is imposed. This reduction in work cannot be achieved by efficient zero checking, a trick used on scalar machines to implicitly
consider symmetry.

To summarize, very high performance is achieved in the transformation if the integral matrix is efficiently squared and a CAL
coded (possibly system supplied) matrix multiply is used. The
use of symmetry represents a major speedup which cannot be obtained
by zero checking. These three basis ideas are the basis for vectorizing the entire CI procedure.

The vectorizing of the CI wavefunction determination has been
independently considered by Saxe and Binkley (6), Saunders and
van Lenthe (7) and this author (8). By far the most complete is
the work of Saunders and van Lenthe and rather than review that
work, we give here a few additional details and refer the reader
to that excellent description of the problem.

The CI Hamiltonian matrix is very sparse; if one attempts to
vectorize the diagonalization of such a matrix, one fully realizes
the importance of the organization of the direct CI techniques.
Not only as a method of eliminating the redundant generation of
formulae and integral indices, but as an organization in which
the regular spacing of both the CI coefficients and the integrals
allows vectorization. In a conventional CI H-matrix, one commonly
has the non-zeros of a given row. To vectorize the contributions
of this row, either the row of H must be expanded to include the
zeros or the CI coefficients, C, and the product of H times C, HC,
which correspond to the non-zeros in the row of H, must be collected. Because of the sparseness of H, the expansion of the row
is not desirable, and one must pack C and HC. The non-zeros of
the row are processed with the packed CI coefficients and added
to the packed HC terms. The packed HC terms must then be scattered
back into the full HC vector. It is this collection and scattering
processes which are so time consuming that the overall performance
of the diagonalization is far from the rate at which the non-zero's
and the collected elements are processed. Clearly, one must avoid
this random collection and scattering if the process is to be made
efficient.

In the direct CI, the CSF's are described as valence, V, if
they have all electrons distributed in the active orbitals (orbitals occupied in any of the reference CSF's), as doublet coupled
single excitations, D, if they have one hole in the active orbital,
and as triplet, T, or singlet, S, coupled double excitations if
they have two electrons outside of the active orbitals. For the
T or S case, a given occupation and coupling of the n-2 electrons
in the active orbitals has associated with it, all possible

distributions of two electrons in the virtual orbitals which lead
to CSF's with appropriate symmetry. In these CSF's, the two electrons in the virtual orbitals are in canonical order, a $\geq$ b. Thus
for each distribution and coupling of the active orbitals, a tail,
we need to know the symmetry of the tail and a pointer to the CI
vector which shows where this group of CSF's begins and we can
generate all the CSF's when needed. Clearly, one does not wish
to have CSF's of the wrong symmetry which must be skipped, but
rather one orders the CSF's blocked by symmetry in a way similar
to the integral order in the transformation. Thus if the symmetry needed by the two electrons was totally symmetric, the CSF's
would order by symmetry aa, bb, cc ... and by canonical order in
each symmetry block.

We consider the simple example of the configurations of one
T or S tail interacting with the other CSF's of the same tail.
Since the CSF's arise from the same tail, all orbitals in the active orbitals must be the same and we need consider only the virtual orbitals. Assume the first CSF has orbitals a and b occupied
and the second, c and d. The matrix element clearly involves the
integrals (ac|bd) and (ad|bc). For an individual CSF with a and
b fixed, we consider the interaction with the other CSF's with c
and d occupied. If we vary c and d, the matrix elements use consecutive elements in the CI vector and contribute to consecutive
elements of the HC vector, as well as requiring the elements associated with the CSF with a and b fixed. If the integrals have been
sorted into the order that for a given a and b, the various c and
d's are present, the integral indexing is very simple. The processing of this block of matrix elements, all use the same formula
(overall coefficients multiplying the integrals); equally important
from vectorization stand point, the elements are in a simple regular location in the vector. In this example, the canonical order
of c-d is the ordering of both the integrals within the integral
block and the CSF's within both C and HC. This regularity of
the integrals, C and HC is possible for all blocks of matrix elements (7). Thus vectorization of all blocks is possible.

With only four different types of tails and a few different
possible types of interactions for some pairs of tails, the number of different blocks is not great and Saunders and van Lenthe
have considered all of them and described how best to treat each
transformation; in the direct CI, the symmetry of a given block
of CI coefficients or a block of integrals can be either totally
symmetric, in which case only a lower triangle is stored, or non-
symmetric, where the matrix is naturally rectangular. The second
difference with the transformation is in some cases when the lower
triangular coefficient matrix is squared into a anti-symmetric
matrix, $C(i,j) = -C(j,i)$. Thus one ends up with a series of cases
in which the integral or the sub-block of CI coefficients must be
squared or both must be squared and many cases where the blocks

are naturally rectangular. The matrix product of the block of
CI coefficients and integrals naturally adds into a block of HC.
In a manner similar to squaring the block of CI coefficients, some
sub-blocks of HC, A = INTEGRALS x CI, must be folded and the A(i,j)
and A(j,i) elements of the product matrix both added to HC(i,j).
In those cases where the A matrix of HC is naturally rectangular,
rather than compute it in a scratch array and then add it into the
total HC, it is better to directly add or subtract the contributions to the existing HC. This is no additional work relative
to a simple MXM where the contributions are added into an array
which has been zeroed out.

To efficiently vectorize a direct CI program, one needs to
design the program with symmetry explicitly included rather than
as an add on. One needs an efficient method of squaring blocks
of integrals and coefficients with the option to change the sign
during the squaring process. An efficient matrix multiply is
needed, preferably one which allows the product matrix to be directly added to, or subtracted from, an existing matrix. One also
needs the ability to fold a matrix, adding both the ij and ji
contributions to a matrix which has been stored as a lower triangle.
The only additional step in the process is to sort the integrals
into the order needed in the CI calculation; this process is
extrinsically vectorizable, by classing the integrals and when a
buffer is full, computing the appropriated sort quantities as a
vector operation. This reduces the sort to a small amount of the
total time and as such becomes unimportant.

REFERENCES

1. H. Lischka, R. Sherpard, F. B. Brown, and I. Shavitt, Intern.
    J. Quantum Chem. S15, 91 (1981).

2. P.E.M. Siegbahn, J. Chem. Phys. 72, 1647 (1980).

3. B. Liu and M. Yoshimine, J. Chem. Phys. 74, 612 (1981).

4. P. Saxe, J. Fox, H. F. Schaefer and N. C. Handy, J. Chem.
    Phys. 77, 5584 (1982).

5. M. Yoshimine, IBM Report, RJ-555, (1969).

6. P. Saxe and J. S. Binkley, private communication.

7. V. R. Saunders and J. H. van Lenthe, Mol. Phys. 48, 923
    (1983).

8. C. W. Bauschlicher, unpublished.

THE METHOD OF SELF CONSISTENT ELECTRON PAIRS. A MATRIX ORIENTED
DIRECT CI

Wilfried Meyer

Fachbereich Chemie
Universität Kaiserslautern
D 6750 Kaiserslautern, Germany

Reinhart Ahlrichs

Institut für Physikalische Chemie und Elektrochemie
Universität Karlsruhe
D 7500 Karlsruhe, Germany

Clifford E. Dykstra

School of Chemical Sciences
University of Illinois
Urbana, Illinois 61801

The self consistent electron pairs (SCEP) method is a matrix
oriented direct CI procedure that is especially well suited to
vectorized computation. The structures and ideas of this method,
along with recent generalizations, are presented. The relation
to other direct correlation approaches and considerations for
vectorization are also discussed.

INTRODUCTION

The remarkable progress which has been made in the theoretical treatment of molecular electronic structures over the last
decade is quite obvious from the size and complexity of the systems which can now be handled, and from the accuracy in energies
and other properties that has been achieved. This progress is
due to the combined effects of refined theoretical models, improved

C. E. Dykstra (ed.),
Advanced Theories and Computational Approaches to the Electronic Structure of Molecules, 19–38.
© 1984 by D. Reidel Publishing Company.

computational algorithms and the increase in power and availability
of computing resources. As the various components of these resources - processors, high speed memory or mass storage - have not
developed uniformly, there has always been a need for adjusting
the methods and algorithms to the computational environment. This
is particularly clear since the advent of new computer architectures based on concepts like pipelining and parallel processing.
Perhaps most striking is the gain in relative speed for processing
arrays of floating point numbers as in algebraic operations such
as scalar products, matrix multiplications etc.

The configuration interaction method (CI) reduces the solution of the N-electron Schrodinger equation to a matrix eigenvalue
problem which should be well suited for vector processing. More
important in this context, however, is the internal structure of
the Hamiltonian matrix elements due to the nature of the basis
elements of the N-electron space, the configuration state functions (CSFs): They are built from antisymmetrized products of
one-electron functions, the molecular orbitals (or spin orbitals),
which temselves constitute a linear vector space. The one-electron
and two-electron integrals over these orbitals, the building
material of the Hamiltonian matrix elements, are tensors of rank
2 and 4, respectively, in this space. Therefore, if care is taken
to construct the CSFs so as to make them tensor components of the
MO space, or rather of its external subspace (as discussed below),
the Hamiltonian matrix elements become tensor components as well
and turn out to be very simple functions of the MO integrals.
The CI expansion coefficients transform contragradiently to the
CSFs, and invariants (tensors of rank zero), like the wavefunction, densities and energies, are then obtained from clean tensor
index contractions between expansion coefficients and CSFs or MO
integrals.

Indeed, all this is quite obvious from explicit expressions
for a spin and space symmetry unrestricted formulation of the CI
where simple Slater determinants are taken as the basis elements
of the N-electron space. But it is clear that, for efficiency,
use should be made of spin and spatial symmetries. Unfortunately,
the tensorial structures are then obscured or lost, if one insists
on orthonormal CSFs, as is usually done. This then results in
complicating the simple index contractions by introducing non-
trivial "coupling coefficients" and thus imparing the vectoriza-
bility of the algorithm. This is particularly a disadvantage for
the direct CI method (1) which must be used when the Hamiltonian
matrix is too large to be stored. The residual vector is then
repeatedly determined from contractions between expansion coefficients and MO integrals.

The self-consistent electron pairs method (SCEP), a type of
direct CI, has been designed to bring out explicitly and to exploit

the tensorial structures in a spin-adapted CI formulation. After
only a rather simple redefinition, the CSFs represent components
of tensors of rank 1 or 2, according to their excitation level.
This leads to very transparent matrix element expressions and
the calculation of the residual vector involves again only tensor
index contractions, i.e. vector and matrix products. We thus
feel that this method is optimally suited for modern array processors. Indeed, the "vectorization" of other direct CI methods,
e.g. those based on the unitary group approach, seems to have led
to the incorporation of some features reminiscent of SCEP.

    The SCEP method was introduced in 1976 by Meyer (2) for
closed shell states and a wavefunction comprising all single and
double substitutions from the reference determinant. Based on
the well defined transformation properties of all pertinent quantities it was demonstrated that one could even depart from the ortho-
normality of the orbitals by allowing for arbitrary linear transformations among the external ones. Thus the formulation given
was actually in an atomic orbital (AO) representation, and such
a form should prove very useful for large molecules. Moreover,
utilizing an idea developed by Ahlrichs and Driessler (3) for
two-electron systems, it was also shown that a full integral transformation could be avoided by calculating two generalized exchange
operators per electron pair directly from the AO integrals. Later
this could even be reduced to a single operator per pair by properly combining singles with pairs (4). First applications were performed by Dykstra, Schaefer and Meyer (5). The method was then
extended to specific higher substitutions by Dykstra (6), and
Cizek's coupled cluster theory was formulated in SCEP terms by
Chiles and Dykstra (7). Starting from an analysis of the structure of the Hamiltonian matrix elements between double substitutions from a multiconfigurational reference wavefunction, given
by Meyer (8), the SCEP method was extended to this general case
by Werner and Reinsch (9), and a similar formulation was independently derived by Ahlrichs (10). A number of recent applications
are discussed in the contribution of Werner and Reinsch to this
volume, in the context of the adaptation of their program to the
CRAY computer. Finally, we note that quite recently a reformulation of the closed shell case in terms of pairwise nonorthogonal
configurations has been given (11), resulting in a further reduction of the computational effort.

    In this contribution we review the SCEP formalism with em-
phasis on the structure of the matrix elements as elucidated by
their transformation properties. We also discuss the use of this
formalism in the coupled cluster approach. This is followed by
a brief discussion of recent computations with a GUGA based program which has been vectorized for the CYBER 205 computer.
torized for the CYBER 205 computer.

SCEP THEORY

1. Transformation Properties of Integrals, Configuration State Functions and Expansion Coefficients

We assume that a reasonable approximation to the true wave-function can be obtained in the form of a linear combination of a limited number of CSFs, provided the latter are built from an optimized subset of molecular orbitals, the "internal" MO's. This reference wavefunction is then augmented by CSFs which have one or two electrons in "external" orbitals which are orthogonal to the internal MO space. All other CSFs have vanishing Hamiltonian matrix elements with the reference function and it is therefore assumed that they are negligible or can be treated as unlinked products of singles and doubles in a coupled cluster ansatz (12) or a coupled electron pair approximation (13). Thus we consider only matrix elements between internal, singly external and doubly external CSFs. If an arbitrary occupation pattern is allowed for in the CSFs of the reference function, the internal space is highly structured and the internal MO's cannot be changed without destroying the compact expansion of the reference wave-function. This structure will be reflected by "internal coupling coefficients" in the matrix elements. In contrast, the external space - which is much larger than the internal space in a calcula-tion with a high quality basis set - is completely unstructured and we should be free to change its basis elements.

Although it is generally useful to think in terms of ortho-gonal orbitals, the transformation properties become clearer if non-unitary transformations are also considered, for which contra-gradient tensors are distinguishable. While the internal orbitals, henceforth denoted by indices i,j,k etc., are required to be ortho-gonal, $\langle \phi_i | \phi_b \rangle = \delta_{ij}$, for the external orbitals, denoted by indices a,b,c etc., we insist only that $\langle \phi_a | \phi_i \rangle = 0$, but allow $\langle \phi_a | \phi_b \rangle = S_{ab} \neq \delta_{ab}$. In order to distinguish covariant and contra-variant transformations we shall consistently use upper and lower indices even if this leads to some deviation from traditional notation. Index contractions for forming lower rank tensors are then only allowed between upper and lower indices and we adopt the convention that summation is implied when an external index appears twice.

Assume now an external orbital transforms according to

$$\phi_a = \tilde{\phi}_b \, t^b_{\ a} \tag{1}$$

under a general linear transformation $t^b_{\ a}$. The overlap integrals $S_{ab} = \langle \phi_a | \phi_b \rangle$ as well as the one-electron integrals $h_{ab} = \langle \phi_a | \hat{h} | \phi_b \rangle$ then transform as

$$h_{ab} = \langle \tilde{\phi}_c | \hat{h} | \tilde{\phi}_d \rangle \, t^c_a {}^* \, t^d_b \; ; \; \underline{h} = \underline{t}^\dagger \, \underline{\tilde{h}} \, \underline{t} \tag{2}$$

(We shall use matrix notation wherever possible in order to emphasize pure index contractions. $\underline{t}$ is the matrix of vectors with
elements t^b_a where a designates a column vector with components
designated by b.) The h_{ab} integrals represent a tensor of rank
two (or a matrix) whereas the elements h_{ia} form tensors of rank
1 (or vectors), i.e. $h_{ia} = \tilde{h}_{ib} \, t^b_a$. The two-electron integrals
$\langle \phi_a \phi_b | 1/r_{12} | \phi_c \phi_d \rangle = (ac|bd)$ transform like tensors of ranks 0 to
4 depending on the number of external indices. For use with the
matrix notation it is convenient to introduce the vectors
$(\underline{I}_{ijk})_a = (ij|ka)$ and the matrices

$$(\underline{J}_{ij})_{ab} = (ab|ij) \; , \quad (\underline{K}_{ij})_{ab} = (ai|kb) \tag{3}$$

They are just ordered sets of two-electron integrals. They obey
$\underline{J}^\dagger_{ij} = \underline{J}_{ij}$, $\underline{K}^\dagger_{ij} = \underline{K}_{ji}$ and transform like $\underline{h}$. It should be noted,
though, that in the case of complex orbitals one has to differen-
tiate between $\underline{K}$ and $\underline{K}'$, $(\underline{K}'_{ij})_{ab} = (ia|kb)$, which transforms like
$\underline{K}' = \underline{t}^T \underline{K} \, \underline{t}$.

The transformation of CSFs and expansion coefficients is
best exemplified for the simple two-electron case. Each of the
following sets of Slater determinants, the spin unrestricted set
$\Psi_{ab} = |\phi_a \phi_b|$ as well as the spin adapted sets,

$$\Psi_{ab} = : \; (|\phi_a \overline{\phi}_b| \pm |\phi_b \overline{\phi}_a|)/2, \; |\phi_a \phi_b|/\sqrt{2}, \; |\overline{\phi}_a \overline{\phi}_b|/\sqrt{2} \; , \tag{4}$$

transform according to

$$\Psi_{ab} = \tilde{\Psi}_{cd} \, t^c_a \, t^d_b \; ; \; \underline{\Psi} = \underline{t}^T \, \underline{\tilde{\Psi}} \, \underline{t} \tag{5}$$

(Note again the difference between this and the transformation
of the one-electron integrals, eq. (2), which is unavoidable for
complex MO's. This difference does not complicate the following
derivations but it causes some notational inconvenience. Since
molecular electronic structure calculations are usually performed
with real orbitals, we may restrict ourselves to considering only
real transformations for which $\underline{t}^T = \underline{t}^\dagger$. In this case $\underline{K}$ and $\underline{K}'$
coincide.) Now, the wavefunction of any physical two-electron
state has to be independent of the orbital set chosen:

$$\Psi = \Psi_{ab} \, C^{ab} = \tilde{\Psi}_{cd} \, \tilde{C}^{cd} = \mathrm{Tr}\,(\underline{\Psi}\,\underline{C}) \tag{6}$$

Substitution using eq. (5) yields the relation between $\tilde{\underline{C}}$ and $\underline{C}$:

$$\Psi = \tilde{\Psi}_{cd} \, t^c{}_a \, t^d{}_b \, C^{ab} \tag{7}$$

$$\tilde{C}^{cd} = C^{ab} \, t^c{}_a \, t^d{}_b \quad ; \quad \tilde{\underline{C}} = \underline{t}\,\underline{C}\,\underline{t}^\dagger \tag{8}$$

The wavefunction has the form of a trace over a tensor product of the CSFs with a tensor of coefficients which transforms contragradient to the CSFs.

It is important to note here that this particular transformation of the coefficient matrix is incompatible with the notion of a nonredundant orthonormal set of spin adapted CSFs. For our spin adapted sets we have

$$\Psi_{ab} = p\,\Psi_{ba} \quad\text{implying}\quad \underline{C}^\dagger = p\,\underline{C}, \; p = \begin{matrix} +1 \text{ singlet} \\ -1 \text{ triplet} \end{matrix} \tag{9}$$

and, assuming for the moment orthogonal orbitals, $\langle \Psi_{ab}|\Psi_{ab}\rangle = (1 + p\delta_{ab})/2$. We note that the above normalization is that used in the work of Meyer (2), Dykstra et al. (5-7) and Werner and Reinsch (9) whereas Ahlrichs (10) and Saunders and van Lenthe (14) have introduced a further factor $1/\sqrt{2}$ in order to make $\Psi_{ab}+p\Psi_{ba} = 2\Psi_{ab}$ normalized for $a \neq b$ so that the off-diagonal expansion coefficients become identical to those of traditional CI.

There is an advantage from the special normalization that becomes obvious from the simple form that results for the overlap and Hamiltonian matrix elements between two arbitrary two-electron wavefunctions:

$$\langle \Psi_1|\Psi_2\rangle = S_{ac}\,S_{bd}\,C_1^{ba*}\,C_2^{dc} = \mathrm{tr}\,(\underline{C}_1^\dagger\,\underline{S}\,\underline{C}_2\,\underline{S}) \tag{10}$$

$$\langle \Psi_1|\hat{H}|\Psi_2\rangle = \mathrm{tr}\,(\underline{C}_1^\dagger\,(\underline{h}\underline{C}_2\underline{S} + \underline{S}\underline{C}_2\underline{h} + \underline{K}[\underline{C}_2])\,) \tag{11}$$

In the latter expression we have introduced the generalized exchange operator

$$K[\underline{C}]_{ab} = (ac|bd)\ c^{cd} \tag{12}$$

for which on finds the transformation

$$\underline{K}[\underline{C}] = \underline{t}^{\dagger}\tilde{\underline{K}}[\tilde{\underline{C}}]\ \underline{t} = \underline{t}^{\dagger}\ \tilde{\underline{K}}[\underline{tCt}^{\dagger}]\ \underline{t} \tag{13}$$

This shows that it can readily be obtained directly from the AO integrals if $\underline{t}$ is extended to a nonsquare matrix (2,3). Eq. (11) is certainly in a form which is optimally suited for array processing.

2. CSFs for N-Electron States

The extension of the above concept to the N-electron case is straightforward. All that is necessary is just to make sure that the CSFs are ordered into groups so that those with n electrons in the external space form tensors of rank n. Their expansion coefficients then form contravariant tensors, again with the result that matrix elements cannot contain coupling coefficients that depend on external indices. Clearly, any particular CSF generates such a group if its external orbitals run over the whole set, i.e. all CSFs of each tensor share the same internal structure. We shall identify such internal structures by indices P,Q,.. for external electron pairs, by S,T.. for single external electrons and by I,J,... for completely internal configurations, including those of the reference wavefunction Ψ_0. The total wavefunction is then written as

$$\Psi = \Sigma_I\ \Psi_I\ c^I + \Sigma_S\ \Psi_{Sa}\ c_S^a + \Sigma_P\ \Psi_{Pab}\ c_P^{ab} \tag{14}$$

The actual construction of the CSFs may start from orthonormal sets of internal CSFs for N-1 and N-2 electrons, respectively, followed by proper spin-coupling with external electrons or electron pairs. A canonical set designed to facilitate the evaluation of coupling coefficients is obtained when using the unitary group approach (15) provided the external MO's are coupled first, as suggested by Brooks et al. (16). It should not be overlooked, though, that one is not interested in the full canonical set but rather in that subset which interacts with the reference function. The smallest set that spans the "interacting space" (17) of a general multi-configurational reference function may involve linear combinations of the canonical CSFs.

A convenient way to construct this smallest set has been described by Meyer (8) using creation and annihilation operators.

Consider those terms of the 2-electron part of the Hamiltonian
for which the two annihilation operators are internal:

$$\hat{H}^{(2)} = \Sigma_{i \geq j} \; (1+\delta_{ij})^{-1} (ri|sj) \; \Sigma_{mn} \; \hat{\eta}_i^{rm\dagger} \; \hat{\eta}^{sn\dagger} \; \hat{\eta}_i^m \; \hat{\eta}_j^n \tag{15}$$

Here, and in similar spin summations given later, it is understood
that $\hat{\eta}^m = 0$ unless $|m| = 1/2$. Since $\hat{H}$ is a "singlet operator",
each term in the brackets, when applied to a spin-adapted reference
CSF Ψ_K, creates a CSF which has the same spin property and a non-
vanishing interaction with Ψ_0. The resulting doubly external
CSFs - r and s are both external in this case - would all be hole-
particle singlet coupled, with the two CSFs $\Psi_{ij,ab}$ and $\Psi_{ij,ba}$
being linearly independent but nonorthogonal. It has recently
been shown by Pulay et al. (11) that this construction is optimal
for a closed shell reference determinant since the number of pairs
is then minimized and there is only one type of them. For general
open shell states it appears, however, that the possibility of a
simultaneous nonorthogonality in the internal and external parts
of the CSFs causes complications with respect to the factorization
of the matrix elements. They are avoided with pair-pair, hole-hole
coupled CSFs which are characteristic for most electron pair
theories, like the independent electron pair approximation, IEPA
(18), the pair natural-orbital CI, PNO-CI (19), various forms of
the coupled electron pair approach, CEPA (13), and the spin-adapted
coupled cluster theory (20). Indeed, $\hat{H}$ can be rearranged into

$$\hat{H}^{(2)} = \Sigma_{i \geq j,p} \; (1+\delta_{ij})^{-1} ((ri|sj) + p(rj|si))$$

$$x \; \Sigma_m \; \hat{B}_{pm}^{rs\dagger} \; \hat{B}_{ij,pm} \tag{16}$$

with the spin-adapted pair creation and annihilation operators
$\hat{B}^{\dagger}$ and $\hat{B}$

$$\hat{B}_{ij,p(m+n)} = (1+\delta_{o,m+n})^{1/2} \; T(pij) \; \hat{\eta}_i^m \; \hat{\eta}_j^n \;, \quad m \geq n \tag{17}$$

Here, $T(pij)$ (anti)symmetrizes the following expression with respect to the indices i and j, i.e. $T(pij)X_{ij} = (X_{ij}+pX_{ji})/2$. The
doubly external CSFs may thus be factored as

$$\Psi_{Pab} = 1/\sqrt{2} \; \Sigma_m \; \hat{B}_{ab,pm}^{\dagger} \; \Phi_{Pm} \tag{18}$$

where the Φ_{Pm} are assumed to form a linearly independent set of
(N-2)-electron internal functions obtained from all the $B_{ij,pm}\Psi_K$.
The overlap integrals between two such CSFs are

$$\langle \Psi_{Pab} | \Psi_{Qcd} \rangle = T(pab)T(qcd)\ S_{ac}S_{bd}\ \Sigma_m \langle \Phi_{Pm} | \Phi_{Qm} \rangle \tag{19}$$

The factors $T(pab)$ just reflect the parity of the CSFs with respect to interchange of their indices. The external factor in
(19) leads to zero for $p \neq q$. Thus the two sets of Φ_{Pm} for $p = \pm 1$
can independently be transformed such that

$$\Sigma_m \langle \Phi_{Pm} | \Phi_{Qm} \rangle \rightarrow \delta_{PQ} \tag{20}$$

For the simple closed-shell case, (20) is immediately valid for
the $\Phi_{Pm} = B_{ij,pm}\Psi_0$ and one can identify the labels $P \equiv ijp$. This
direct link to the annihilated orbitals is lost in the general
case after the transformation to arrive at (20), but P is always
understood to contain the parity p. Singly external CSFs may
similarly be obtained from $\Sigma_m \hat{B}^\dagger_{ak,pm} \hat{B}_{ij,pm} \Psi_K$ and are written as

$$\Psi_{Sa} = \Sigma_m \hat{n}_a^{m\dagger} \Phi_{Sm} \tag{21}$$

The internal functions Φ_{Sm} may again be linearly combined to
yield $\Sigma_m \langle \Phi_{Sm} | \Phi_{Tm} \rangle = \delta_{S,T}$ so that $\langle \Psi_{Sa} | \Psi_{Tb} \rangle = S_{ab}$.

    We conclude this section by noting that the structures of
the CSFs given in (18) and (21) are valid independently of the
method used to obtain them, e.g., GUGA with the external electrons
coupled first. These structures will be used in the discussion
of the Hamiltonian matrix elements where the explicit factorization of the CSF's into external and internal parts allows simple
explicit expressions for the internal coupling coefficients.
The particular advantage of the scheme outlined here is that it
opens the way to an "internal contraction" of the CSFs by applying the $\hat{B}_{ij,pm}$ to the reference function as a whole instead of
to its individual CSFs. This substantially reduces the number of
independent pair functions, and consequently the computing effort,
with only minor losses in energy. This is discussed in more detail in this volume in the contribution of Werner and Reinsch.

3. Hamiltonian Matrix Elements and Residual Vectors

    The structure of the matrix elements between any two of the
CSFs discussed above is easily derived from the requirement that
they be linear in the integrals with the correct tensorial form

obtained by appropriate factors of the overlap matrix. As out-
lined, we have available a single rank 4 tensor, $(ab|cd)$, a single
type of rank 3 tensors, $(ia|bc)$, but several types of tensors of
rank two or lower from $\underline{h}$, $\underline{J}$, $\underline{K}$ and $\underline{S}$. The tensorial structure of
the matrix elements allows any linear combination among them. It
is therefore useful to introduce as the general tensor of rank 2
the following Fock type operator:

$$\underline{F}(A,B) = \underline{h}\ \sigma(A,B) + \underline{S}\ \gamma(A,B)$$

$$+ \Sigma_{kl}\ [\underline{J}_{kl}\ \alpha_{lk}(A,B) - \underline{K}_{kl}\ \beta_{lk}(A,B)] \tag{22}$$

The labels A and B may have the actual values I,P,S and denote
the internal functions Ψ_I, Φ_{Pm}, and Φ_{Sm}, respectively. Other
possible values are "klI," denoting $B_{kl,pm}\Psi_I$, and "jS," denoting
the internal function defined from

$$\Sigma_m\ \eta_a^{m\dagger}\ \Phi_{jSm} = \Sigma_m\ B_{aj,pm}^{\dagger}\ \Phi_{Pm}$$

They determine completely the reduced coupling coefficients
σ,α,β and γ, which correspond to Siegbahn's internal parts (21)
"B" factored out of the total coupling coefficients "A". (The
notation used here for the coupling coefficients is reminiscent
of Roothaan's definition of the open-shell Fock operator (22).)
The two indices of F must originate from different sides of the
matrix elements, as is clear from the components $\underline{h}$, $\underline{J}$ and $\underline{S}$. Two
indices from the same CSF can only appear with $\underline{K}'$. The general
tensor of rank 1 can be written as a row of $\underline{F}$ with a fixed internal
index: $\underline{F}(A,B)_j$. The matrix elements are consequently obtained
as

$$\langle\Psi_I|\hat{H}|\Psi_J\rangle \quad = \gamma(I,J) \tag{23}$$

$$\langle\Psi_{Sa}|\hat{H}|\Psi_I\rangle \quad = \Sigma_j\ F(S,jI)_{aj} \tag{24}$$

$$\langle\Psi_{Sa}|\hat{H}|\Psi_{Tb}\rangle \quad = F(S,T)_{ab} \tag{25}$$

$$\langle\Psi_{Pab}|\hat{H}|\Psi_I\rangle \quad = T(pab)\ \Sigma_{kl}\ (K_{kl})_{ab}\ \sigma(P,klI)/\sqrt{2} \tag{26}$$

$$\langle\Psi_{Pab}|\hat{H}|\Psi_{Sc}\rangle = T(pab)\ \Sigma_j\ [(ac|bj)\ \sigma(P,jS)$$

$$+ F(P,jS)_{aj}\ S_{bd}]/\sqrt{2} \tag{27}$$

$$\langle \Psi_{Pab} | \hat{H} | \Psi_{Qcd} \rangle = T(pab)T(qcd)$$

$$\times [(ac|bd)\ \sigma(P,Q) + 2F(P,Q)_{ac}\ S_{bd}] \qquad (28)$$

It is worth noting here that these equations can be subjected to
partial transformations in the sense that only a subset of the tensor indices are involved, e.g. they are valid even if different sets
of orbitals are used for different pairs. This opens the way for
a pair natural orbital treatment (19,23). Indeed, the underlying reduction of the coupling coefficients was first demonstrated
in this context (8). The above set of equations is complete with
the following relations for the coupling coefficients which can
be derived (9) from eqs. (15), (18) and (21) using the well known
η operator algebra. (Note the difference, however, between

$$[\eta^a, \eta_b^\dagger] = \delta^a_b \quad \text{and} \quad [\eta_a, \eta_b^\dagger] = S_{ab}$$

which leads, e.g., to eq. (19)).

$$\sigma(A,B) = \sum_m \langle \Phi_{Am} | \Phi_{Bm} \rangle \qquad (29)$$

$$\alpha_{kl}(A,B) = \sum_m \langle \Phi_{Am} | \sum_n \hat{\eta}_k^{n\dagger}\ \hat{\eta}_l^n | \Phi_{Bm} \rangle \qquad (30)$$

$$\beta_{kl}(A,B) = \sum_{mm'} \langle \Phi_{Am} | \sum_n \hat{\eta}_k^{(m-n)\dagger}\ \hat{\eta}_l^{(m'-n)} | \Phi_{Bm'} \rangle)$$

$$\times\ [(1+\delta_{om})(1+\delta_{om'})]^{-1/2} \qquad (31)$$

$$\gamma(A,B) = \sum_m \langle \Phi_{Am} | \hat{H}^{(N-Nex)} | \Phi_{Bm} \rangle / Nex \qquad (32)$$

Due to overlap integrals $\langle a|j \rangle$ in the external part of the matrix
elements, β_{kl} and γ have to be set to zero if A or B are jS or jI.
Instead of actually using these expressions, which are based on
the factorization of the CSFs, one may apply standard procedures
for evaluating the matrix elements of eqs. (23-28) for specific
sets of external indices and then extracting the reduced coupling
coefficients. This is the basic idea of the symbolic matrix
method of Liu (24). The number of non-zero coupling coefficients
is strongly dependent on the complexity of the reference wavefunction. In the closed-shell case they reduce to very simple expressions (2), e.g.

$$\alpha_{kl}(ijp,mnq) = \delta_{pq} \, T(pij) \, T(qmn) \, \delta_{ki}\delta_{lm}\delta_{jn}$$

$$x \, [(1+\delta_{ij})(1+\delta_{mn})]^{-1/2} \tag{33}$$

In a direct CI the iterative solution of the eigenvalue equation proceeds via the residual wavefunction, which is the projection of $|\hat{H} - E|\Psi\rangle$ onto the configuration space. The computationally crucial step is therefore the contraction of the matrix elements with expansion coefficients to form the following G tensors:

$$\langle\Psi_{Pab}|\hat{H}|\Psi\rangle = T(pab) \, (G_P)_{ab}$$

$$\langle\Psi_{Sa}|\hat{H}|\Psi\rangle = (G_S)_a$$

$$\langle\Psi_I|\hat{H}|\Psi\rangle = G_I \tag{34}$$

$$\underline{G}_P = \Sigma_Q [\underline{K}[\underline{C}_Q] \, \sigma(P,Q) + 2\underline{F}(P,Q) \, \underline{C}_Q \, \underline{S}]$$

$$+ \Sigma_S \, \Sigma_j \, \sqrt{2}\underline{F}(P,jS)_j \, \underline{C}_S^\dagger \, S + \underline{K}[\underline{C}_P']$$

$$+ \Sigma_I \, \Sigma_{kl} \, \underline{K}_{kl} \, \sigma(P,klI)/\sqrt{2} \, C_I \tag{35}$$

$$\underline{G}_S = \Sigma_P \, \Sigma_j \, P[\underline{K}[\underline{C}_P]]_j \, \sigma(P,jS) + \underline{S} \, \underline{C}_P \, \underline{F}(P,jS)_j]\sqrt{2}$$

$$+ \Sigma_T \, \underline{F}(S,T) \, \underline{C}_T + \Sigma_I \, \Sigma_j \, \underline{F}(S,jI)_j \, C_I \tag{36}$$

$$G_I = \Sigma_P \, \Sigma_{kl} \, tr(\underline{C}_P^\dagger \, \underline{K}_{kl}) \, \sigma(P,klI)/\sqrt{2}$$

$$+ \Sigma_S \, \Sigma_j \, \underline{F}(S,jI)_j^\dagger \, \underline{C}_S + \Sigma_J \, \gamma(I,J) \, C_J \tag{37}$$

The coefficient matrix $\underline{C}_P'$ in eq. (35) is defined as $C_{Paj}' = 2^{-1/2} \Sigma_S \, C_S^a \, \sigma(P,jS)$. $\underline{C}_P$ and $\underline{C}_P'$ can be added up so that only one $\underline{K}[\underline{C}]$ per pair is required. Eqs. (35-37) demonstrate most clearly the absence of any external coupling coefficients as the direct consequence of the transformation properties of the CSFs.

The iterative updating of the expansion coefficients $\underline{C}$ → $\underline{C} + \underline{\Delta}$, asks for the approximate solution of

$$\sum_x \langle \Psi_x | \hat{H}-E | \Psi_y \rangle \Delta^y = -G_x + E \sum_y \langle \Psi_x | \Psi_y \rangle c^y \tag{38}$$

where x and y stand for I, Sa and Pab, respectively. For simple approximations of $\langle \Psi_x | \hat{H}-E | \Psi_y \rangle$ and ease of inversion it is at this point, if not done from the beginning, useful to transform to orthogonal orbitals, $S_{ab} \to \delta_{ab}$, and orthogonal configurations, $\sigma(P,Q) \to \delta_{PQ}$. It is then sufficient to assume that

$$\langle \Psi_x | \hat{H}-E | \Psi_y \rangle \approx \delta_{xy} E_x$$

Different sets of orthogonal orbitals may be chosen for different pairs in order to obtain an optimal approximation (2), which is particularly helpful in the case of localized internal orbitals. In the context of the internally contracted SCEP it may be advantageous to work up to this point with nonorthogonal internal functions just as they are obtained from applying the $B_{ij,pm}$ to Ψ_0 since they have a more compact form and the lists of the coupling coefficients are shorter.

Finally, we remark that the F operators in eqs. (35-37) may either be calculated once and stored or else be recalculated as they are needed. This depends on the storage and control processor capability. A detailed discussion of the optimal sequences for performing the various summations appearing in eqs. (35-37) has been given in references 9 and 14.

4.   Operators for Coupled Cluster Wavefunctions

The coupled cluster approach (CCA), as discussed in detail elsewhere in this volume, builds in higher order correlation effects by means of an exponentially expanded wavefunction $\Psi_{CCA} = \exp(\hat{S})\Psi_0$. $\hat{S}$ is a substitution operator involving one $(\hat{S}_1)$, two $(\hat{S}_2)$, and up to N electrons. $\hat{S}_2$ can be given by a sum of spin-adapted pair creation and annihilation operators as in eq. (16). The CCA wavefunction is found by projection of the CSFs included in $\hat{S}$ onto $|\hat{H}-E|\psi\rangle$, and so it is suited to a direct, iterative solution. In fact, the requisite Hamiltonian matrix elements differ from those of the correspondingly substituted CI only by the interaction of the CI's CSFs with the new substitutions from the third and higher power terms of the series expansion of $e^{\hat{S}}$. Of course, after a few terms, such interactions involve differences of more than two spin orbitals and so are zero.

The most important higher order correlations accounted for with CCA are obtained when $\hat{S}$ is limited to $\hat{S}_2$ (or $\hat{S}_1+\hat{S}_2$), yielding the double substitution CC wavefunction designated CCD. Chiles and Dykstra (7) derived a direct, matrix oriented method for obtaining CCD wavefunctions. In the closed shell case, we can, as already mentioned, identify P = ijp, Q = klq and so forth. Also, $\Sigma_Q \to \Sigma_{k>1} \Sigma_{q=\pm1}$. If for convenience, the coefficient matrices are scaled acoording to $C'_{ijp} = C_{ijp}(1+\delta_{ij})^{1/2}$ the operator to be added to equation (35) is:

$$G^{CCD}_{\underline{-}ijp} = [2/(1+\delta_{ij})]^{1/2} \, T(ijp) \, \underline{S} \, \Sigma_{kl} \tag{39}$$

$$\{ \, (2-p)^{-1/2} \, tr(\underline{C}'_{ijp} \, \underline{K}_{kl}) \, \underline{C}'_{klp}$$

$$- \Sigma_q \, (2-q)^{1/2} \, [\underline{C}'_{ijp}\underline{K}_{kl}\underline{C}'_{klq} + \underline{K}[\underline{C}'_{ikq}]_{lk} \, \underline{C}'_{ljp}]$$

$$+ \frac{1}{2} \Sigma_{q,r} \, \underline{C}'_{ikq} \, [\Omega(pqr) \, \underline{K}_{kl}$$

$$+ \Theta(pqr) \, \underline{K}_{lk}] \, \underline{C}'_{ljr} \, \} \, \underline{S}$$

The functions Ω and Θ are simple internal spin factors given by the values of p, q and r:

$$\Omega(pqr) = [\, (2-p) \, (2-q) \, (2-r) \,]^{1/2}$$

and

$$\Theta(pqr) = (p + pqr - (2-r)^{1/2} \, (2-q)^{1/2}) \, / \, (2-p)^{1/2}$$

(There are other equivalent expressions for Θ (7) that have been given.)

MATRIX FORMULATED GUGA MR-CI (SD)

The COLUMBUS program system for electronic structure calculations, developed by F. Brown, H. Lischka, R. Pitzer, I. Shavitt
and R. Shepard has been adapted for the CYBER 205 in Karlsruhe.
This has been carried out by R. Ahlrichs, H. J. Boehm, C. Ehrhardt,
P. Scharf, H. Schiffer, H. Lischka and M. Schindler. The programs
fully exploit symmetry of D_{2h} and its subgroups. Inspection re-
vealed that the CI code would not be vectorizable on the CYBER 205
to any appreciable extent. Loops involved index incrementation
larger than one - vectorizable on the CRAY-1 but not on the
CYBER 205 - and, what is more serious, the code included nested
loops with the constraints $i>j>k$, which should be avoided if
possible to reduce the number of start ups for vector instructions.
However, the program structure was still well adapted for a matrix
formulation: The (transformed) integrals and CI coefficients were
(essentially) ordered as matrices $\underline{J}_{ij}$, $\underline{K}_{ij}$ and $\underline{C}_p$, and the reduced
coupling coefficients of the GUGA, i.e. the internal parts, are,
of course, just what is needed for a matrix formulated code,
namely the α,β,γ for the canonical CSF's.

Since the processing of the two-external integrals, i.e.
the $\underline{J}_{ij}$ and $\underline{K}_{ij}$ matrices, usually requires most of the time, this
part was rewritten first. Since this step could be formulated in
terms of matrix multiplications this was simple in principle.
The CI coefficients had to be expanded into full matrices $\underline{C}_p$, and
the linear combinations of $\underline{J}$ and $\underline{K}$ had to be arranged in proper
matrix form. The resulting contributions to the residual vector
were then also obtained as full matrices and had to be brought
back into the usual form in which (anti) symmetric matrices are
normally stored. For the singlet case, p=1, this required, of
course, the appropriate modification of diagonal elements C_p^{aa}
(before) and G_{Paa} (after) the matrix multiplication by a factor
$\sqrt{2}$. With these modifications the entire two-external part then
reduced to a simple matrix multiplication for symmetry blocked
matrices. Since the diagonal elements of the Hamiltonian usually
involve larger sums of integral contributions they are constructed
once and for all in the COLUMBUS program and processed independently. This requires some rather trivial modifications of the
processing of J and K matrices to avoid certain contributions
being counted twice.

These program developments were done on a scalar computer.
The changes considerably reduced the length of the code and reduced the machine code produced by the FORTRAN compiler to 30%.
The CPU time was virtually unchanged (±5%). Since the original
code was carefully designed (no logics in the innermost loops)
the matrix code gains about as much through better loop structure
as it loses in the additional operations required to extend and
later on to (anti) symmetrize matrices.

We next tackled the processing of four-external integrals, i.e. the construction of the terms described in eq. (12). This development is still in progress. In the present version we construct the combinations of integrals

$$(ac,bd)_\pm = [(ab|cd) \pm (ad|bc) \ (1+\delta_{ac})(1 + \delta_{bd})]^{-1/2}$$

which guarantees that no logic is required later on if all CSF's are normalized to unity. Since we wanted to avoid a complete reordering, the present order of the integrals leads to a vector instruction with length L, the dimension of externals of a given symmetry; only in 1/3 of cases the vector length is L^2. The most important parts concerning the one external and all internal integrals were easily vectorized. We have not yet rewritten the processing of three external integrals, but this is under way. It appears that the CYBER 205 is not too well suited for tasks like matrix multiplication (or similarly our present processing of four externals):

Loop i

Loop j

vectorized loop: $A(k,i) = A(k,i) + B(k,j) \ C(j,i)$

If the indices i,j,k run from 1 to N, a realistic timing for a single pipe machine is

$$T = N^2 \ (130 + N) \ \text{cycles}$$

(1 cycle = 20 nsec. For n pipes replace 130 + N by 130 + N/n). The start up time for the kernel, a triadic operation, is 80 cycles but an additional 50 cycles are required to prepare the vector instruction, e.g. loading of registers, and other tasks which we have not been able to reduce so far. This leads to an asymptotic speed, V_∞ = 100 MFLOPS as N → ∞, but half the asymptotic speed is reached for $N_{1/2} \approx 130$. In the CI(SD) one typically has matrix multiplications of dimension 10 to 70 resulting in a performance of 7 to 35 MFLOPS. The CRAY-1 performs matrix multiplications with (24) $V_\infty \gtrsim 130$ MFLOPS and $N_{1/2} \approx 7$ i.e. one easily gets a performance $\gtrsim 100$ MFLOPS, roughly a factor of 5 better than for the CYBER 205. The state of affairs is much better for the integral transformation, where the transformation itself can in fact be done at ≥ 50 MFLOPS on a single pipe CYBER 205.

In the following table we compare timings of the program as it stands now with those given by Werner and Reinsch in this volume and those of Saunders and van Lenthe (14).

Table I. Comparison of SD-CI(SD) Calculations for Ozone[a]

	$^1A_1$			$^3B_2$		
	SCEP	Ref. 14	present	SCEP	Ref. 14	present
N_{CI}[b]	35279	31440	31440	41482	223393	37166
CPU-times:[c]						
Transformation	3.4	49.	33.	3.6	49.	33.
Per iteration	3.3	2.4	9.2	4.7	4.1	15.0
Number of iterations	11.	11.	8.	7.	13.	8.
Total	39.7	75.	107.	32.8	144.	155.
$-E_{SCF}$	--	224.32737	224.32737	--	224.39623	224.39623
$-E_{corr}$	0.58195	0.58194	0.58194	0.50417	0.50494	0.50416

[a] Basis: [641], 69CGTO's. Only valence MOs were correlated and the three high lying virtual orbitals were excluded. (See ref. 14 and the contribution of Werner and Reinsch in this volume.)
[b] Number of CSF's included. For the $^3B_2$ state, the present calculation is done with the interacting space restriction.
[c] In seconds. SCEP times and times from ref. 14 are for a CRAY-1. The present calculation times are for a CYBER 205.
[d] The convergence limit in the present calculation was 10^{-8} a.u. in the energy while in ref. 14 it was 10^{-5} a.u. for the CI vector.

The iteration times for the $^1A_1$ state demonstrate mainly the shortcomings of the CYBER 205 which is about a factor 3-4 slower for this kind of computation, in agreement with the comments made above. The Karlsruhe integral transformation appears to be slightly faster than the SvL (14) CRAY-1 program. More than 40% of the Karlsruhe transformation program is required for sorting of integrals (not vectorized) and other tasks such as construction of the core Hamiltonian. We note that this also applies for the $(H_2O)_2$ timings of SvL. O_3 is certainly an unfavorable case for the CYBER 205 since the number of externals (for $^1A_1$) are 21,17, 1C,6 in the corresponding symmetries and start up times involve a large fraction of the CPU time. The results for the $^3B_2$ state mainly show that little is lost (0.1% of the correlation energy) by imposing the interacting space restriction. It is also obvious that freezing high-lying virtuals has virtually no effect on the correlation energy.

SUMMARY

The method of self-consistent pairs, since its introduction, has been a powerful approach to the electron correlation problem because of how nicely it exploits the tensorial structures in spin-adapted direct CI. This has meant that it was well suited to the minicomputers of a decade ago that had small amounts of memory but no virtual memory software, since the matrices used by SCEP were never larger than the Fock matrix. Now, with the development of supercomputers and array processors, SCEP continues to be well suited because the computationally important steps are immediately vectorizable. As we have discussed here, SCEP has evolved into a method which is general with respect to the reference state function and is even applicable to the coupled cluster approach.

REFERENCES

1.   B. Roos, Chem. Phys. Lett. 15, 153 (1972).

2.   W. Meyer, J. Chem. Phys. 64, 2901 (1976).

3.   R. Ahlrichs and F. Driessler, Theor. Chim. Acta 36, 275 (1975).

4.   W. Meyer in "Post Hartree-Fock: Configuration Interaction" ed. W. A. Lester (Publication LBL-8233, University of California, Berkeley, 1978); R. Ahlrichs, Comp. Phys. Comm. 17, 31 (1979).

5. C. E. Dykstra, H. F. Schaefer and W. Meyer, J. Chem. Phys. 65, 2740 (1976); J. Chem. Phys. 65, 5141 (1976).

6. C. E. Dykstra, J. Chem. Phys. 72, 2928 (1980).

7. R. A. Chiles and C. E. Dykstra, J. Chem. Phys. 74, 4544 (1981).

8. W. Meyer in Vol. 3, "Modern Theoretical Chemistry," H. F. Schaefer (Plenum, New York, 1977).

9. H.-J. Werner and E. A. Reinsch, in "Proceedings of the 5th Seminar on Computational Methods in Quantum Chemistry," eds. T. H. van Duinen and W. C. Niewpoort, (MPI Garching, Muenchen, 1981); J. Chem. Phys. 76, 3144 (1982).

10. R. Ahlrichs, in "Proceedings of the 5th Seminar on Computational Methods in Quantum Chemistry," eds. T. H. van Duinen and W. C. Niewpoort, (MPI Garching, Muenchen, 1981); R. Ahlrichs, in "Methods in Computational Molecular Physics," eds. G.H.F. Diercksen and S. Wilson (Reidel, Dordrecht, Holland, 1983).

11. P. Pulay, S. Saebo and W. Meyer, J. Chem. Phys. 81, 000 (1984).

12. J. Cizek, J. Chem. Phys. 45, 4256 (1966); B. G. Adams and J. Paldus, Phys. Rev. B20, 1 (1979).

13. W. Meyer, J. Chem. Phys. 58, 1017 (1983).

14. V. R. Saunders and J. H. van Lenthe, Mol. Phys. 48, 923 (1983).

15. J. Paldus, J. Chem. Phys. 61, 5321 (1974); Phys. Rev. A14, 1620 (1976); I. Shavitt, Int. J. Quant. Chem. Symp. 11, 131 (1977).

16. B. R. Brooks and H. F. Schaefer III, J. Chem. Phys. 70, 5092 (1979); B. R. Brooks, W. D. Laidig, P. Saxe, N. C. Handy and H. F. Schaefer III, Physica Scripta 21, 312 (1980).

17. A. D. McLean and B. Liu, J. Chem. Phys. 58, 1066 (1973).

18. M. Jungen and R. Ahlrichs, Theor. Chim. Acta (Berlin) 17, 339 (1970).

19. W. Meyer, Int. J. Quantum Chem. S5, 341 (1971); R. Ahlrichs, H. Lischka, V. Staemmler and W. Kutzelnigg, J. Chem. Phys. 62, 1225 (1975).

20. J. Paldus, P. G. Adams and J. Cizek, Int. J. Quant. Chem.
     11, 813 (1977).

21. P.E.M. Siegbahn, J. Chem. Phys. 72, 1647 (1980).

22. C.C.J. Roothaan, Rev. Mod. Phys. 32, 179 (1960).

23. P. R. Taylor, J. Chem. Phys. 74, 1256 (1981).

24. R. W. Hockney, C. R. Jesshope, "Parallel Computers" (Adam
     Hilger Ltd., Bristol, 1981).

EVALUATION AND PROCESSING OF INTEGRALS

Dermot Hegarty

Dept. of Chemical Physics, University of Groningen,
Nijenborgh 16, 9747 AG Groningen, The Netherlands

INTRODUCTION

The vast majority of computational methods used in quantum
chemistry can be completely described as evaluation and processing
of integrals. Expectation values are written as a non-linear combination of integrals where the coefficients describe the wavefunction. Methods for calculating the wavefunction from the
variational principle applied to the energy expectation value may
be many and varied, but usually reduce to a prescription for the
processing of integrals.

In this paper we obviously cannot discuss the whole of quantum
chemistry and thus we restrict to the evaluation and processing
of integrals over atomic orbitals, that is, over functions describing (either well or poorly) free atoms. We also only discuss
the popular methods based on the LCAO model for molecular orbitals.

Our primary concern is the computational aspects of methods
and how vector processing affects the development. Since computational quantum chemical methods and generally available computing
facilities have grown side by side ever since the early sixties it
would be naive to suggest that these methods were anything other
than tightly bound to computer architectures. Consequently, a
dramatic change in architecture (such as vector processing but
much less so small scale parallel processing) should be expected
to produce a change in efficient computational methods although
the induced change may be just a change in emphasis and not philosophy.

The integrals that occur in quantum chemistry are either one-

C. E. Dykstra (ed.),
Advanced Theories and Computational Approaches to the Electronic Structure of Molecules, 39–66.
© 1984 by D. Reidel Publishing Company.

or two-particle integrals, i.e. over the three dimensions of one
particle or over six dimensions of two particles. it is the latter
type in which we are exclusively interested. While these need not
be any more complicated than the 1-particle type, there are many
more of them and it is their number that causes the most difficulties in otherwise simple algorithms.

We begin by considering the numerical calculation, moving on
to the classes of processing required in wavefunction determination methods. Also included is a section on symmetry, developing
previous works (11-14). We assume throughout that the reader has
a working knowledge of SCF, MCSCF and CI methods together with at
least an idea of the difference between scalar and vector processors. Considerable emphasis is placed on a (2-pipe) CYBER 205
computer system, a machine with which we have most experience,
with relevant comments where necessary about the CRAY system.

EVALUATION OF 2-ELECTRON INTEGRALS OVER GAUSSIAN FUNCTIONS

Gaussian functions are by far the most popular basis functions
used in ab initio quantum chemistry. This is due to the possible
reduction of 4-center, 6-dimensional two electron integrals to an
analytic form involving just a 1-dimensional integral (1) which
can be evaluated relatively easily. contracted sets of functions
(i.e. linear combinations with fixed coefficients) are universally
used to counter the inherent misbehaviour of single gaussians
giving the basis a better form without increasing the number of
variational parameters.

The calculation of gaussian integrals must be performed by
blocks: a block contains that set of integrals available from a
set of 4 shells and each shell contains a set of components. For
example, the [dp|ps] block involves a d shell with 6 (or 5) compo-
nents, p with 3 and s with 1, thus a total of 54 (or 45) integrals.
Blocks play an important role because many of the intermediate
quantities calculated can be used for many integrals and thus do
not need recalculation. Fig. 1(a) shows the dependence of the
theoretically required number of arithmetic (floating point) operations per uncontracted integral within a block (5), for what
are probably the two most efficient general algorithms. The dotted
line is an algorithm developed by McMurchie and Davidson (2),
based on the properties of Hermite polynomials, and the solid line
a development by Saunders (3) of an algorithm due to Dupuis, Rys
and King (4) based on an exact numerical quadrature method.

Particularly noteworthy is the almost linear behaviour of
Saunders' algorithm as the quantum numbers increase illustrating
the importance of calculating integrals in blocks. Also noteworthy
is the sharp increase in cost at low quantum numbers, this is due

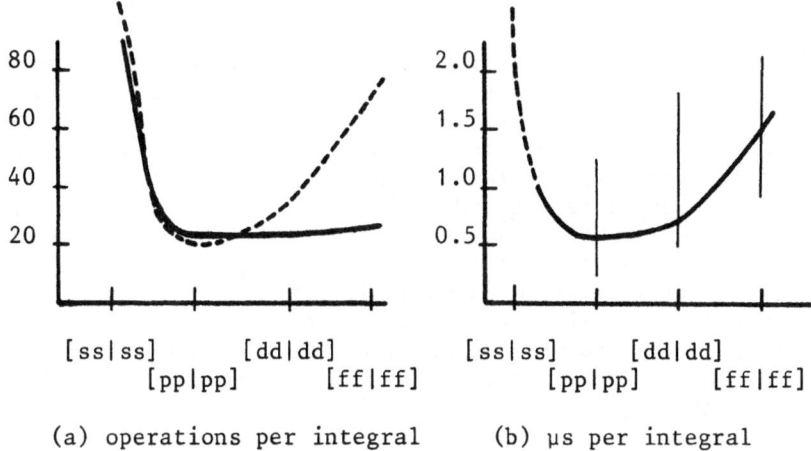

(a) operations per integral   (b) μs per integral

Fig. 1   Prediction and Practise in Gaussian Integral Evaluation

to the large, quantum-number-independent cost of each block that
is borne by a single integral in an [ss|ss] block but divided over
a possible 81 in a [pp|pp] block. This also illustrates the need
to calculate integrals in blocks, the [ps|ss] blocks will be ~3
times faster because of division of this initial preparatory cost.
A detailed discussion of the steps involved in these two algorithms
is to be found in ref. (5) and the outcome of this study was the
development (6) of an efficient vectorized program for the Cyber
205 computer, following McMurchie's and Davidson's scheme. The
actual performance of this program is shown in Fig. 1(b) (again
for uncontracted integrals). The shape corresponds to that in
Fig. 1(a) but the [ss|ss] block is now relatively much slower due
to the use of the scalar processor in setting up blocks, the high
cost of the scalar processor is felt much more strongly by the
[ss|ss] blocks than any other. In this respect the program can be
improved for low quantum numbers by removing even this trivial use
of the scalar processor (improvement is factor of 2 for [ss|ss],
thus ~2/3 for [ps|ss] and ~1/40 for [pp|pp] etc.). This also il-
lustrates a general point concerning the balance between use of
scalar and vector processors within a given computer.

     Having determined upon an efficient method, on the basis of
the necessary number of arithmetic operations (3,5) and the likely
values of parameters affecting this, we can ascertain how the
architecture of vector processors can be best utilized. Fig. 2
lists, on the left, 3 computers with pipe-lined vector capabilities,
and on the right 3 possible approaches to program development.
The first of the latter relies on a detailed knowledge of the
computer so that a program may be written so that all arithmetic
units and memory channels are busy at the same time, thus ensuring

high efficiency. Considering the sophistication of modern architectures this is not a recommended course, and indeed, as the arrows
in Fig. 2 from right to left are intended to indicate, only the
FPS164 (and all scalar processors) are reasonably handled in this
way. A much better approach is to develop the algorithm so that
as much emphasis as possible is placed on standard macro operations
instead of (apparently) unrelated primitive operations. Thus, a
matrix multiplication is a standard macro operation, it consists
of n^2 inner products which can again be considered standard macro
operations, and these in turn toal to $n^2(2n-1)$ primitive operations. Matrix multiplication is uniquely defined, thus is is reasonable to expect compilers to recognize this program construct
and to generate code to use the machine as efficiently as possible.
On the other hand, an arbitrary collection of $n^2(2n-1)$ primitive
operations remains just that and efficiency is a question of luck.

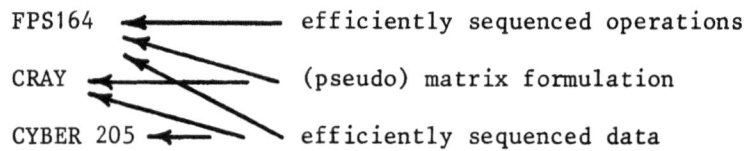

Fig. 2 Computer Architectures and Program Design

     The kernel of the McMurchie and Davidson algorithm can be
written as a pseudo, sparse, double matrix multiplication (3),

$$G_{abcd} = \underline{\underline{D}}^{ab} \, \underline{\underline{R}}^{abcd} \, \underline{\underline{D}}^{cd} \qquad (1)$$

(we have taken some considerable liberty with notation; it is not
our intention to rederive expressions but only to illustrate these),
where G_{abcd} represents a 2-electron integral. A program along
these lines is expected to reflect much more closely the mathematical development. Note that only the concept of macro operation
has been used; in order to use a specific computer architecture
efficiently we also need to know which subset of the infinite
number of defineable macro operations can use the architecture
efficiently. In the case of vector processors this is the usual
set of vector operations (from which the above matrix product can
be constructed) and possibly some logical (as opposed to arithmetic)
equivalents or combinations of the two (for example, data gathering
and scattering and the use of bit vectors on the CYBER 205).

     Unfortunately the vector reformultation of quantum chemical
methods still leaves us with a high degree of machine dependence,
the CRAY will operate efficiently if offered vectors of length
40-64 whereas the CYBER 205 does not reach 50% efficiency until
the length reaches ~150. Thus if we strive for matrix and vector

operations as the essential content then we must ensure that the
order is sufficiently large. The CYBER 205 is much less suited to
general chemical problems where the natural order is often <100.
The order in eq. (1) above does not reach acceptable levels even
for the CRAY unless high quantum numbers are involved and this
leads us to the third approach to program design. This last option
relies on our knowledge that normal calculations involve many more
than one block of integrals and those of the same quantum numbers
require (almost) exactly the same treatment. A parallel approach
is a natural suggestion and that vector processor hardware is
serial pipelined rather than parallel is irrelevant. This approach
is obviously of great importance in quantum chemistry since computational methods are dominated by repetition. Consider the construction of a closed shell Fock matrix, each 2-electron integral makes
its contribution independently and thus all can be processed together. In practise this is oversimplified because data movement
is often more expensive than arithmetic operations and a matrix
reformulation is better suited to minimize this. However in the
case of the calculation of integrals the parallel approach is
ideal, we calculate a single integral from one block in parallel
with a similar integral (same directional quantum numbers) from a
second block, in parallel with a third etc. Vectorization is almost
complete, whatever is done to one integral must be done to all and
thus every operation is a vector operation. This is important
since even a small contribution from the slow scalar processor
can seriously degrade efficiency (more seriously on the CYBER 205
than the CRAY that has a relatively fast scalar processor). However, since parallel sets of data are required, the memory requirement is correspondingly larger. This places an upper limit on the
lengths of vectors but in our CYBER 205 implementation even [ff|ff]
blocks (intermediate storage increases with quantum number) permit
vector lengths >150 in $\frac{1}{2}$Mword of memory. The vertical bars in Fig.
1(b) indicate the timing range that can be expected by varying
lengths of vectors, the lower limits correspond to >10000 and the
upper limits to ~100, the measured performance is obtained by fil-
ling $\frac{1}{2}$Mword.

Note that all blocks forming the vector should have the same
characteristics, including equivalences of individual integrals
due to permutation symmetry. This restriction is undesirable since
it increases program complexity and reduces the maximum vector
lengths available to a basis set. The problem can be partially
solved by ordering the blocks appearing in a vector as shown in
Fig. 3.

The entire vector, A, is used where permutational symmetry
is not relevant, and the subvector B where contributions to a
single integral $[a_i b_j | c_k d_l]$ (a_i is the ith component of shell a)
are to be calculated and $k < l$, and A where $k > l$. For basis sets
of reasonable size the problem is not significant since the number

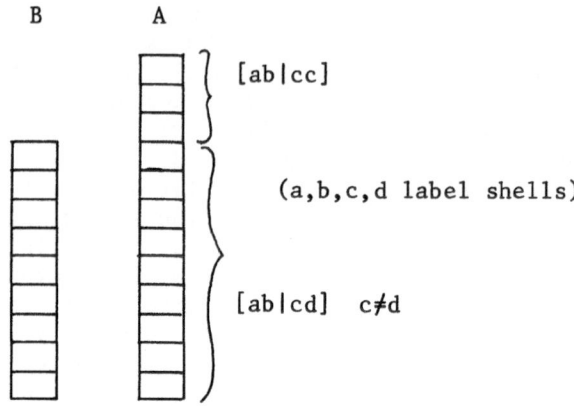

Fig. 3   Ordering of Blocks to Maximize Permutation Symmetry

of blocks with a=b, or c=d, or a=c and b=d is relatively small and
thus we can safely calculate all component integrals and delete
those that are permutationally redundant (easily done using bit
vectors on the CYBER 205).

An obvious advantage in using vectors of blocks is that there
is no need to consider special cases, e.g. [ss|ss] or [ps|ss],
since the time wasting steps (e.g. loops of length 1) are divided
over all elements of the vector, thus this becomes negligible.
Quite generally, vector processors reduce the need for efficient
scalar code if full use of the vector processor is made. One class
of special case deserves special mention, that of 1- and some 2-
center integrals. In these cases a trivial algebraic reduction of
the integral formulae can be made that leads to considerable savings in number of operations. Where this is important (diatomic
molecules and possibly complexes where the central atom is described by a large basis) we must use a technique similar to that
in Fig. 3 or ensure that all blocks in a vector are identical in
this respect.

It should also be realised that vector processors do not
have the full complement of vector operations implemented in hardware, a well known omission in the CRAY is that of scatter/gather
data movement operations. While these must be considered vector
operations they must be implemented using the scalar processor.
An operation that is missing in the hardware of both CRAY and
CYBER 205 is that of contraction depicted schematically in Fig. 4.
The elements grouped by braces are to be arithmetically combined
to give a single result, for example, combination may be the average, or the sum or product etc. In the contraction of integrals

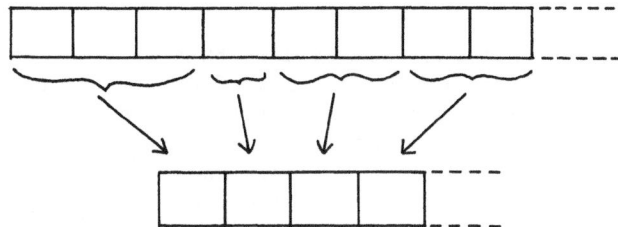

Fig. 4   Schematic Representation of Contraction Operation

the combination is the sum (assuming the vector multiplication
by contraction coefficients has already been performed). If each
set of grouped elements is considered a subvector, and the result
element a scalar quantity then both the CRAY and CYBER 205 have
suitable vector operations, however the efficiency is very low if
the lengths of these subvectors is <5. The McMurchie-Davidson
algorithm contracts over charge distributions so that the number
of elements is often 1, sometimes 2-5 and occassionally 4-25.
Saunders' algorithm contracts over primitive integrals so that
these lengths become 1,2-5, 4-25, 8-125, 16-625 with the lower
numbers occuring most often. We have already considered restricting
blocks in a vector to those containing the same permutational
symmetry and the same number of centers and now we could add the
same degree of contraction. This then introduces a regularity into
Fig. 4 that can be exploited. However this places extremely stingent
constraints on possible vectors and alternative solutions are
preferable. Such a solution can be found on the CYBER 205 if bit
vectors are used, and also has more diverse applications e.g.
sparse matrix by vector product where only non-zero elements are
stored with labels. A further solution is to use the fact that
all components of a shell are equally contracted so that all component charge distributions and integrals are too, thus if all 81
vectors are available for a given [pp|pp] vector of blocks then
contraction can vectorize across the components instead of down
the vector of a single component. However problems with short
vectors remain for [ss|ss], [ps|ss] blocks and for [dd|dd] blocks
the memory requirement becomes excessive and thus this solution
is not generally applicable.

One final point will be made concerning contraction; Fig. 1(à)
shows that the minimum number of operations required to calculate
integrals can be as low as ~20, each additional operation (e.g.
label calculation, threshold comparison etc.) then adds ~5%. Let
us suppose that contraction is given to the scalar processor running 20 times slower, the single addition operation becomes 20
effective operation times and the total time is doubled. While
contraction is trivial on scalar-only computers it has become

noteworthy on computers possessing vector capabilities.

To summarize what has been said in this section: integrals over gaussian functions can be calculated extremely efficiently by blocks and extremely simply on vector processors by taking vectors of blocks as long as on is aware of:
a) memory requirement and its influence on vector length
b) permutational symmetry within a block can be handled dynamically
c) the one-center special case is not always worthwhile but should be handled statically
d) contraction is not trivial and any additional operations must be considered carefully.

PROCESSING OF 2-ELECTRON INTEGRALS

We will first consider some general aspects of the processing of integrals and in a later section the specific processing required in SCF, MCSCF and CI methods.

Firstly, the order in which gaussian integrals are calculated is not what could be considered a standard order. The order is predetermined (by the block structure of vectors) but will appear more random than ordered. Fig. 5 illustrates the order, Fig. 5(a)

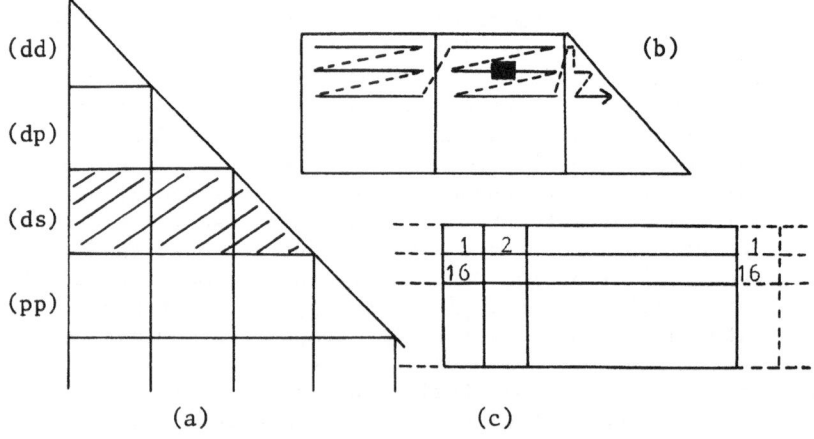

Fig. 5 Order of Calculated Integrals

show the integral supermatrix blocked by the quantum numbers of charge distributions. Integrals will be calculated from left to right and top to bottom. The shaded line of Fig. 5(a) is expanded in 5(b) to show the internal order within a quantum number group. The solid line is followed until the arrow head, the row in which it occurs being determined by the necessary memory for intermediate data, the dotted lines show the continuations. The single block indicated in Fig. 5(b) is expanded in 5(c) to show the component integrals. A vector of integrals labelled '1' will first appear sequentially, than a vector of those labelled '2' etc. Thus if we require the submatrix G_{ij} (the set of permutationally inequivalent integrals [ij|kl] for fixed functions i,j) then we must be prepared to find the elements interspersed with the elements of, at least, the supermatrix G_{IJ} where I,J label the shells to which i,j respectively belong. The CYBER 205 requires much longer vectors to operate efficiently and thus the chance that several G_{IJ}'s are also interspersed is consequently greater. Note that we place the (dd) charge distribution in the corner of the triangular matrix of Fig. 4(a); this helps to minimize the number of elements within G_{ij} where the number of G_{ij} within G_{IJ} is largest, and of course should f functions be contained in the basis then these would preceed the d's.

Should we be content with a reordering of calculated integrals then the above discussion is irrelevant, however reordering is random data movement and is a slow operation. A list in main memory can be ordered in one movement operation per element running at ~8 and ~15 Mmoves per second for the CRAY and CYBER 205 respectively, compared to standard vector arithmetic times of 80 and ~80 MFLOPS. If the list is too large for main memory (almost always the case with 2-electron integrals) then it must be first reduced to smaller chunks. Using a bucket sort each pass of this phase can probably still run at ~8 Mmoves per sec on the CRAY but the CYBER 205 performance will certainly not be better than ~5 Mmoves and these results will only be obtained with very carefully written machine code for the scalar processors. A binary sort can use the vector processor of the CYBER 205 (with bit vectors) and can thus be written clearly in a high level language but the number of passes increases much more rapidly than with the bucket sort (the binary sort is a 2-bucket sort), and while each pass is more efficient it rapidly becomes less efficient. A second important factor is the IO wait time in ordering large lists. The number of operations performed on each data item transferred from external storage is too few so that the processor must constantly wait. In a multiprocessing system it should switch to another task but for the bucket sort to be efficient most of main memory must be used so that there cannot be any other tasks. In this respect the binary sort is preferable.

Even if we neglect IO problems concerned with ordering we should not expect the scalar ordering step to be significantly faster than the vector calculation step. An estimate can be found for a 2-pipe CYBER 205 if we assume 60 buckets, 1Mword main memory, a bucket sort time of ~8 cycles (an optimistically estimated machine code routine assuming no conflicts within processor or memory and no overheads) and a final scatter of 4 cycles per integral.

Let us further assume that only non-zero integrals are stored.
For basis sets up to ~100 functions the cost is 12 cycles or 24
arithmetic operation times per integral, up to ~300 functions the
cost is 40 arithmetic operation times, above ~300 it becomes 56.
Fig. 1(b) suggests that even a simple program can calculate integrals and labels in an average 0.5 μs or 50 arithmetic times (con-
vergence to theoretical predictions does not occur until [dd|dd]
blocks) so that ordering will not be any faster than a factor of
2 and probably much less. This assumes an uncontracted basis,
with a normal contraction scheme and using the McMurchie-Davidson
algorithm (2) this factor must be multiplied by $\frac{1}{2}(n_p/n_c)^4$ with n_p
primitive and n_c contracted functions.

It should be obvious that any development that allows quantum
chemical systems to be described by fewer basis functions will
have an immediate effect on the processing of integrals. The use
of Slater (exponential) type functions rather than gaussians in
atoms leads to such a reduction although the general case of molecules is much less clear. There are clearly identifiable advantages
in using Slater functions, densities at nucleii are better described
and the long range behaviour of wavefunctions is improved. However,
the description of chemical bonding is not heavily influenced by
either of these aspects and thus there is no a priori reason to
suppose that fewer Slater functions would be required. Despite
this the majority opinion is in favour of Slater functions and
the only real objection is the costly calculation of multicenter
integrals. Progress is continuously being made in numerical methods
(7) and we quote here a development by Michels (8) for multicenter
integrals that is suited to vector processors. A spherical harmonic
expansion is used to shift centers of charge distributions followed
by numerical integration. The time required for the calculation
of integrals over N basis functions of the same quantum numbers
increases as

$$T = aN + bN^2 + cN^4 \quad \text{where } a \gg b \gg c \qquad (2)$$

and the constants a, b depend on the quantum numbers, but not c,
and are such that the N^4 term only becomes dominant at large $N(\approx 50)$.
Even so the timing (8) appears to be 10^2-10^3 times slower than
our implementation (6) of the McMurchie-Davidson algorithm and
thus we conclude that the use of Slater functions is still not
quite competitive with gaussians.

In order to improve the atomic character of basis functions
it would appear better to use the general contraction scheme of
Raffenetti (9) with gaussian functions and extend the atomic basis
with additional uncontracted gaussians to provide the flexibility
needed in molecules. This has been suggested by Ruedenberg et al.
(10) as a good starting point for MCSCF calculations. The general
contraction scheme is however more like a small blocked transforma-

tion than a contraction, the implementation of which is far more expensive than segmented contraction.

The use of spherical gaussian functions (i.e. adapted to R(3) symmetry) also reduces the number of basis functions if d and higher are included in the basis (5 or 6 d functions leads to 625 or 1296 integrals in a [dd|dd] block, a considerable difference). The transformation to spherical functions is rather easier to implement than general contraction since the transformation is local to a block although the number of integrals required simultaneously may reduce the possible vector lengths (only a problem on the CYBER 205). In any case it is better to calculate integrals directly over spherical functions if the algorithm allows, and this is indeed the case with the McMurchie-Davidson algorithm and at no extra cost.

Point group symmetry can considerably reduce the number of non-zero integrals by either blocking the 2-electron integral supermatrix by choosing as basis symmetry adapted orbitals, or taking advantage of equivalences among groups of integrals induced by symmetry operators. The first method is the most popular and requires a small transformation since integrals must be evaluated over cartesian functions. If a transformation to MO's is required in any case then the intermediate transformation to symmetry orbitals (SO's) should be performed since this blocks the full transformation. Davidson (14) has analysed methods for doing this in some detail.

One additional advantage to using SO's as basis is that matrix representations of other operators become blocked and thus less main memory is required although this no longer has the importance that it once did. The operations required to construct these operators from a symmetry blocked 2-electron integral list are certainly fewer than if symmetry were not exploited, but not less then if the second method mentioned above was used (and possibly more if multidimensional representations are involved).

The method was suggested by Dacre (11), and later developed and extended by several authors (12,13), for the special case of symmetry in closed shell SCF calculations. The basis is the AO basis but all the advantages of an SO basis are obtained (except the blocking of Fock and density matrices) without requiring an explicit transformation to SO's. In the next section we generalize the method in rather a simple way to include generalised coulomb and exchange matrices, and thus to MCSCF and CI methods.

SYMMETRY AND SYMMETRY OPERATORS

The unitary operator $\hat{R}$ transforms the orthogonal MO basis $\{\phi_i\}$

$$\hat{R}\,\phi_i = \sum_j \mathcal{R}_{ij}\phi_j \tag{3}$$

and the same operator transforms the AO basis $\{\psi_a\}$

$$\hat{R}\,\psi_a = \sum_b R_{ab}\psi_b \tag{4}$$

Thus $\mathcal{R}$ and $\underline{\underline{R}}$ are completely defined. The MO basis is related to the AO basis through

$$\phi_i = \sum_a c_{ia}\psi_a \tag{5}$$

and thus

$$\hat{R}\,\phi_i = \sum_a c_{ia}\hat{R}\psi_a \tag{6}$$

$$\therefore \sum_b (\sum_j \mathcal{R}_{ij}c_{jb})\psi_b = \sum_b (\sum_a c_{ia}R_{ab})\psi_b \tag{7}$$

If the effect of $\hat{R}$ does not depend on the explicit form of $\{\psi_a\}$ then

$$\mathcal{R}\,\underline{\underline{c}} = \underline{\underline{c}}\,\underline{\underline{R}} \tag{8}$$

and $\quad \underline{\underline{c}} = \mathcal{R}^+\,\underline{\underline{c}}\,\underline{\underline{R}} \tag{9}$

Any unitary operator $\hat{R}$ satisfying this condition is a symmetry operator. Let us now assume the representations of R, and the basis functions $\{\phi_i\}$, $\{\psi_a\}$ are real then

$$\underline{\underline{h}} = \underline{\underline{R}}\,\underline{\underline{h}}\,\tilde{\underline{\underline{R}}} \quad \text{and} \quad \underline{\underline{G}} = \tilde{\underline{\underline{R}}}^{(2)}\underline{\underline{G}}\,\underline{\underline{R}}^{(2)} \tag{10}$$

where $\underline{\underline{h}}$ is the matrix of 1-electron integrals and $\underline{\underline{G}}$ the supermatrix of 2-electron integrals, $\underline{\underline{R}}^{(2)}$ is the direct product representation $\underline{\underline{R}} \times \underline{\underline{R}}$. Both $\underline{\underline{h}}$ and $\underline{\underline{G}}$ are considered square and supersquare matrices.

The relations in eq. (10) can be considered necessary conditions for symmetry operators (they are also satisfied by many non-symmetry operators), or as 'generating' relations. Thus

$$\underline{\underline{h}} = \sum_R \underline{\underline{R}}\,\bar{\underline{\underline{h}}}\,\tilde{\underline{\underline{R}}} \quad \text{and} \quad \underline{\underline{G}} = \sum_R \underline{\underline{R}}^{(2)}\,\bar{\underline{\underline{G}}}\,\tilde{\underline{\underline{R}}}^{(2)} \tag{11}$$

where $\bar{\underline{\underline{h}}}$ and $\bar{\underline{\underline{G}}}$ are sparse matrices where only integrals unique under $\{E,R\}$ are included divided by an appropriate integer to take account of multiple generations.

Both the full, $\underline{\underline{h}}$, $\underline{\underline{G}}$ and skeleton matrices $\bar{\underline{\underline{h}}}$, $\bar{\underline{\underline{G}}}$ have trivial permutational symmetries and in practise we only consider permutationally unique elements. Thus we must consider

$$\underline{\underline{h}} = \underline{\underline{h}}^{(1)} + \underline{\underline{h}}^{(2)} = \sum_R \underline{\underline{R}} \; (\bar{\underline{\underline{h}}}^{(1)} + \bar{\underline{\underline{h}}}^{(2)}) \; \tilde{\underline{\underline{R}}} \tag{12}$$

and $\quad \underline{\underline{G}} = \sum_{i=1}^{8} \underline{\underline{G}}^{(i)} = \underline{\underline{R}}^{(2)} (\sum_{i=1}^{9} \bar{\underline{\underline{G}}}^{(i)}) \; \tilde{\underline{\underline{R}}}^{(2)} \tag{13}$

where the superscripts denote permutationally inequivalent sets of elements of $\underline{\underline{h}}$ and $\underline{\underline{G}}$ and in general

$$\underline{\underline{G}}^{(i)} \neq \sum_R \underline{\underline{R}}^{(2)} \bar{\underline{\underline{G}}}^{(i)} \underline{\underline{R}}^{(2)} \tag{14}$$

The usual manipulations of 2-electron integrals involve contractions with products of coefficients, thus we also need the transformation properties of the direct product $\underline{\underline{c}}^{(2)} = \underline{\underline{c}} \times \underline{\underline{c}}$. From eq. (9) we easily obtain

$$\underline{\underline{c}}^{(2)} = \underline{\mathcal{R}}^{+(2)} \; \underline{\underline{c}}^{(2)} \underline{R}^{(2)} \quad \text{or} \quad \underline{\mathcal{R}}^{(2)} \underline{\underline{c}}^{(2)} = \underline{\underline{c}}^{(2)} \underline{\underline{R}}^{(2)} \tag{15}$$

The coulomb contraction

$$J^{ij}_{\mu\nu} = \sum_{\lambda\sigma} G_{\mu\nu\lambda\sigma} \; c_{i\lambda} \; c_{j\sigma} \tag{16}$$

can be written in matrix form

$$\underline{\underline{J}} = \underline{\underline{G}} \; \tilde{\underline{\underline{c}}}^{(2)} \tag{17}$$

thus from eq. (11)

$$\underline{\underline{J}} = \sum_R \underline{\underline{R}}^{(2)} \bar{\underline{\underline{G}}} \; \tilde{\underline{\underline{c}}}^{(2)} \tilde{\underline{\underline{\mathcal{R}}}}^{(2)} \tag{18}$$

The same expression is valid for $\underline{\underline{K}}$, the exchange contraction, since the ordering of the (super)basis of the $\underline{\underline{G}}$ supermatrix has not been fixed. The skeleton $\bar{\underline{\underline{J}}}$ can be calculated with symmetry unique integrals and expanded to $\underline{\underline{J}}$ with two (non-symmetric) highly blocked transformations.

The coulomb and exchange contributions to the closed shell Fock matrix is a special case, the elements of $\underline{\underline{c}}^{(2)}$ are contracted to give the one-electron density matrix $\underline{\underline{D}}$. The unitarity of $\underline{\underline{\mathcal{R}}}$ ensures the transformation property

$$\underline{\underline{D}} = \tilde{\underline{\underline{R}}} \; \underline{\underline{D}} \; \underline{\underline{R}} \quad \text{or} \quad \underline{\underline{D}}^{(2)} = \underline{\underline{D}}^{(2)} \underline{\underline{R}}^{(2)}$$

giving

$$\underline{\underline{G}} \; \underline{\widetilde{D}}^{(2)} = \sum_R \underline{\underline{R}}^{(2)} \underline{\underline{G}} \; \underline{\widetilde{D}}^{(2)} = \sum_R \underline{\underline{R}} [\underline{\underline{G}} \; \underline{\widetilde{D}}^{(2)}] \underline{\underline{R}}$$

where the square brackets denote a change of structure, from super-vector to matrix.

INTEGRAL PROCESSING IN SCF, MCSCF AND CI METHODS

The single determinant SCF method involves the construction of Fock matrices directly from integrals and a subsequent diagonalization. For example the closed shell Fock matrix is given by

$$F_{\mu\nu} = h_{\mu\nu} + \sum_{\lambda\sigma} D_{\lambda\sigma} (2G_{\mu\nu\lambda\sigma} - G_{\mu\lambda\nu\sigma}) \qquad (19)$$

The construction of $\underline{\underline{F}}$ is then an n^4 process, the diagonalization better than n^3 and the construction of $\underline{\underline{D}}$ is mn^2 where n is the number of AO's and m the number of occupied orbitals. It is clear that integral processing is the dominant step.

In MCSCF calculations there are three important processes, a partial transformation, the solution of the MCSCF equations and the diagonalization of the many electron hamiltonian. With conventional multiconfiguration model wavefunctions the last part is small whereas for CAS type models it is the second part that is small. The ratios are of the order 70:20:10 (see examples in (15), (16) and 50:0:50 (examples in (17), (18)) although variations can be quite large especially for CAS models where the CI time increases very rapidly with the number of valence orbitals, the transformation time more slowly. It is reasonable to expect that integral processing, in at least the partial transformation will not be insignificant.

In CI methods the distribution of effort is over the transformation, construction of 'formula tape' and diagonalization. Direct CI is characterized by the very short 'formula tapes' required whereas this is a significant step in conventional CI (H matrix element driven). MO integral driven direct CI differs from AO integral driven in that the latter replaces the full n^5 transformation with a partial mn^4 transformation and an iterative m^2n^4 contraction. Conventional CI differs from the direct versions by calculating the hamiltonian matrix explicitly whereas the direct methods calculate the contributions to $\sigma = \underline{\underline{H}}.\underline{c}$ directly by multiplying integrals with the appropriate elements of $\underline{c}$. The contribution to processing times from transformations depends heavily on the molecular system and accuracy required in the calculation so that integral processing may or may not be significant.

Thus we see that AO integral processing become less important

as the accuracy of calculations increases, but that it only becomes negligible for very highly accurate calculations, i.e. where
the number of variational parameters is comparable with the number
of integrals.

Let us now consider the manipulations of integrals that must
be performed in each of these methods and how this can be related
to vector processing.

Without any doubt the cheapest (in floating point operations)
approach to the construction of the closed Fock matrix $\underset{\approx}{F}$ is via
the $\underset{\approx}{P}$ supermatrix defined as

$$P_{\mu\nu\lambda\sigma} = 4G_{\mu\nu\lambda\sigma} - G_{\mu\lambda\nu\sigma} - G_{\mu\sigma\nu\lambda} \qquad (20)$$

which exhibits the same symmetries as $\underset{\approx}{G}$. We consider a basis of
permutationally inequivalent charge distributions so that $\underset{\approx}{P}$, $\underset{\approx}{G}$
are lower triangular and of order $n(n+1)/2$, $\underset{\approx}{F}$ is written as a
(super)vector of the same order. Rows of $\underset{\approx}{P}$ are denoted $\underline{P}^{(\mu\nu)}$ and
elements of $\underline{F}$ as $F_{(\mu\nu)}$, then

1a) $F_{(\mu\nu)} = \underline{P}^{(\mu\nu)} \cdot \underline{D}$ dot product of $[\mu\nu]$ elements

1b) $\underline{F} \quad := \underline{F} + \underline{P}^{(\mu\nu)}D_{(\mu\nu)}$ vector increment of $[\mu\nu]$ elements

for all $(\mu\nu)$ in increasing order. Part (a) is the contribution
from $\underline{P}$ and part (b) that from $\underset{\sim}{\underline{P}}$ to the $\mu\nu$th and all previous elements of $\underline{F}$. We use the square bracket notation $[ab]$ to denote
$a(a-1)/2+b$. Thus $4[[n]]$ arithmetic operations which the CYBER 205
can perform in $3[[n]]$ operation times.

In comparison the integral at a time method is considerably
more expensive.

$$
\left.
\begin{array}{ll}
2a) & F^{\mu\nu} := F^{\mu\nu} + 4D_{\lambda\sigma} \\
2b) & F^{\mu\lambda} := F^{\mu\lambda} - D_{\nu\sigma} \\
2c) & F^{\mu\sigma} := F^{\mu\sigma} - D_{\nu\lambda} \\
2d) & F^{\mu\sigma} := F^{\mu\sigma} - D_{\nu\lambda} \\
2e) & F^{\nu\lambda} := F^{\nu\lambda} - D^{\mu\sigma} \\
2f) & F^{\nu\sigma}_{\sigma\lambda} := F^{\nu\sigma}_{\sigma\lambda} + 4D^{\lambda}_{\mu\nu}
\end{array}
\right\} \times G_{\mu\nu\lambda\sigma} \quad (\mu \geq \nu,\ \lambda \geq \sigma,\ [\mu\nu] \geq [\lambda\sigma])
$$

Here there are 14 floating point operations per integral. On a
scientific mini computer with fast indexing operations relative
to floating point operations this method will be ~4 times slower.
However, in that case $\underset{\approx}{G}$ does not need to be in any particular
order and thus full use can be made of near-zero integrals and
symmetry according to the previous section without introducing
extra costs.

The formulation for a vector processor needs to be slightly
improved, in an obvious notation we can rewrite case (b) above

as an example

3b) $F_{\mu\lambda} := F_{\mu\lambda} - \underline{\underline{G}}_{\lambda}^{\mu\nu} \cdot \underline{D}_{\lambda}$ dot product of length
λ ($\mu \neq \lambda$) or ν ($\mu = \lambda$)

The vectors are now of the order of the basis functions
(except cases 2a and 2f) whereas the supermatrix formulation
gives the order of charge distributions. This will have a much
smaller effect on the CRAY than the CYBER 205 where it accounts
for a factor ~2 unless n > 200 when it slowly reduces to 1. So
far the supermatrix formulation is ~6 times faster (the supermatrix timing can be accurately predicted, the integral at a time
method has an unpredictably slow scalar component for array indexing).

Given the previously discussed calculated order of $\underline{\underline{G}}$ it is
obvious that the calculation of $\underline{P}$ is dominated by an out-of-memory
sort, and as also discussed an optimistic estimate is >40 operation
times per non-zero integral on the CYBER 205. The sort step is
thus equivalent to >20 iterations. The one integral at a time
method will only require an in-memory ordering.

Neither of the methods can make efficient use of near-zero
integrals since both require structured matrices unless there
are very many zero integrals (see later in this section). The
use of point group symmetry according to the previous section
is similar to the case of non-zero integrals. If SO's are used
then the matrices retain their structure and so can be used efficiently in the supermatrix method, some efficiency is lost however
in the second method since vector lengths are reduced still further (the constant vector startup overhead begins to dominate).

The ordered structure of either $\underline{\underline{G}}$ or $\underline{P}$ is very important
to efficient implementation on vector processors since array in-
dexing and data movement steps are minimized. Let us consider
G as a randomly ordered vector of integrals $\underline{V}$ with appropriate
labels, and consider the parallel approach to the calculation
of contributions to $\underline{F}$. Firstly we need to calculate 6 array in-
deces for all elements of $\underline{V}$ (in practise this vector would be
chopped into smaller units), then gather the appropriate density
matrix elements to $\underline{U}$ in the order required by $\underline{V}$, perform the vector
multiplication $\underline{UV}$, gather the Fock matrix elements and add to
$\underline{UV}$ and finally replace the result in $\underline{F}$. There are two problems
with this approach, firstly and most important is the superproblem
and secondly the process is dominated by non-floating point operations. Even if the superproblem did not exist neither the CRAY
nor CYBER 205 would perform better than 1/20 of floating point
efficiency. This is clearly unsatisfactory.

The superproblem is depicted schematically in Fig. 6 and is

a generalization of the problem encountered earlier with the contraction of integrals. In that problem adjacent elements were
to be arithmetically combined, here the elements occur randomly
throughout the vector. In other words contributions to F_{ij} will
occur randomly through $\underline{UV}$ unless $\underline{V}$ is structured.

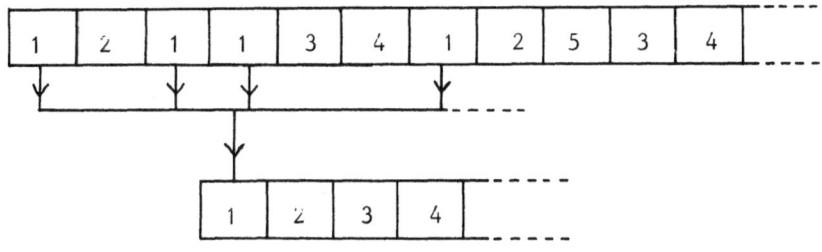

Fig. 6 The Super Problem

When it is extremely important to make use of the sparsity
of $\underline{\underline{P}}$ then we must use a compressed but partially ordered $\underline{\underline{P}}$ in

$$4a) \quad F_{\mu\nu} := F_{\mu\nu} + D_{\lambda\sigma} \Big\}$$
$$4b) \quad F_{\lambda\sigma} := F_{\lambda\sigma} + D_{\mu\nu} \Big\} \times P_{\lambda\sigma}^{(\mu\nu)}$$

and consider the vector to be a randomly ordered sparse $P^{(\mu\nu)}$.
This requires a gather of the elements of $\underline{D}$ in 4a and a gather +
scatter of the elements of $\underline{F}$ in 4b. Each of these data movements
costs 8 operation times on the CYBER 205 (somewhat greater on
the CRAY) which immediately swamps the floating point operations.
The number of non-zero elements must be <11% for this to be worthwhile. Similar data movements can be performed in the integral
at a time method, the relative performances remaining a factor
3 in operation times with an additional factor 1-2 reflecting
the shorter vector lengths when using the CYBER 205.

There are two instances when these latest suggestions become
optimal, firstly in spatially large molecules with high symmetry
(e.g. hydrocarbon chains) and secondly where the number of integrals is so large (large basis sets) that the SCF algorithm must
qualitatively change. This later case is the direct SCF suggestion
of Almlöf et al. (19) wherein integrals are recalculated at each
iteration. The qualitative difference arises through the fact
that quite large integrals can give rise to negligible contributions to Fock matrix elements but in order to take advantage of
this the density matrix must be known before the integral calculation.

It is usual to test the expected magnitude of an integral
(using an approximate expression) against a threshold before the
full calculation is performed and the test expression can be chosen
to be the same for all integrals in a block. Thus a single test
can eliminate an entire block. Almlöf et al. (19) discuss ways
in which the test criteria can incorporate the eventual contribution to the Fock matrix, the simplest follows immediately from
2a - 2f, namely

$$\tau = \max(4|D_{\mu\nu}|, \ |D_{\nu\sigma}|, \ |D_{\nu\lambda}|, \ |D_{\mu\sigma}|, \ |D_{\mu\lambda}|, \ 4|D_{\mu\nu}|) \quad (21)$$

and τ times the test integral is compared to the threshold. Since
integrals are calculated in blocks a compressed density (20) is
required

$$D_{pq} = \max_{\mu\epsilon p, \nu\epsilon q} D_{\mu\nu} \quad (22)$$

where p, q denote shells and μ, ν components.

Contributions to the total energy (20) can also be used to
determine test criteria although Almlöf et al. (19) report convergence difficulties in this case.

In closed shell problems only rotations between occupied
and unoccupied orbitals affect the total energy and convergence
behaviour. Using the contribution of integrals to the off-diagonal
elements of $\underline{\underline{F}}$ in the MO basis we can derive (19) the criteria

$$\tau = \max(4R_{\mu\nu\lambda\sigma}, \ R_{\mu\lambda\nu\sigma}, \ R_{\mu\sigma\nu\lambda}) \quad (23)$$

where

$$R_{\mu\nu\lambda\sigma} = (f_\mu g_\nu + f_\nu g_\mu)|D_{\lambda\sigma}| + (f_\lambda g_\sigma + f_\sigma g_\lambda)|D_{\mu\nu}| \quad (24a)$$

and

$$f_\mu = \max_i |c_{i\mu}| \qquad g_\mu = \max_a |c_{a\mu}| \quad (24b)$$

where $c_{i\mu}$ is a coefficient of the ith occupied MO, and $c_{a\mu}$ the
ath unoccupied MO. This gives a more stringent test than eq.
(21) so eliminating more integral blocks. A further development
recognizes that only the change to the Fock matrix needs to be
computed

$$\underline{\underline{F}}^{(m)} = \underline{\underline{F}}^{(m-1)} + \underline{\underline{P}} \ \Delta\underline{D}^{(m-1)} \quad (25)$$

and $\Delta\underline{D}^{(m-1)}$ is the change in density matrix from iteration (m-1)
to (m-2). This will obviously be numerically smaller leading to
an even more stringent test condition.

Let us now consider the relative costs of conventional and direct methods. We assume that the total number of contracted basis functions is 2/3 of the uncontracted primitives and the average uncontracted integral calculation time is 30 operation times (absolutely efficient routines are essential), giving an average of ~125 operation times per contracted integral. The cost of the supermatrix method (1a,b) on the CYBER 205 is

$(125 + 56)S + 3.I$ with S the sparsity factor due to near-zeroness and I the number of iterations,

and we have included 56 operation times (optimistic) for the (large basis) sort.

Direct SCF can use the integral at a time method (3b) or the supermatrix method (4a,b) where the latter is three times faster but requires that integrals be calculated three times as often (to avoid sorting). The former costs

$(125 + 27)S'.I$ with S' the total sparsity (i.e. only the required integrals).

When the condition

$$\frac{181}{152} \cdot \frac{S}{I} > S'-2 \quad \text{with S,S' expressed in percent,}$$

is satisfied then the direct method is to be prefered. This is always the case when $S' \approx 2\%$ and Table 1 shows the relationship between I and S'/x when S = 50% and 75%, where x is the efficiency ratio of conventional to direct. The table shows that it is not unreasonable to expect that the direct method will often be competitive (x=1) and even where it is slower (x>1) it would be surprising to find a large factor. An obvious exception occurs with open shell calculations where many states are calculated using the same $\underline{\underline{P}}$. In addition IO is completely avoided, an aspect that has not been taken into consideration.

Table 1   Maximum number of Iterations for which Direct SCF is prefered

S'/x	S = 50%	S = 75%
50	1	1
25	2	3
10	7	11
5	19	29
2	∞	∞

We have considered the closed shell SCF method in some detail since while it is simple, it illustrates the problems that are encountered with vector processors (it is slightly misleading to call these 'problems', the only real problem is to surpress one's outdated concepts of computers and efficient algorithms).

The processing of integrals that we will consider in MCSCF and CI methods revolve around 4-index transformations and 3-index contractions. This latter occurs in direct MCSCF and direct direct-CI, these bear the same relations to MCSCF and direct CI as direct SCF bears to conventional SCF.

A recent detailed examination of transformation methods has been given by Saunders and Van Lenthe (21) who make optimal use of matrix algebra not only to define an efficient algorithm but to reduce the cost relative to previously reported methods (22). Fig. 7 shows the algorithm and it is strikingly obvious that it fits well on vector processors. However, some comments are necessary. The sparsity of the matrices should be used, at least in the most time consuming steps. For a complete transformation this is the step labelled (a) in Fig. 7, if $\underline{\underline{c}}$ is non-square with ratio $<2^{-\frac{1}{2}}$ then the step labelled (b) becomes the most expensive. The outer product algorithm is suitable:

suppose $\underline{\underline{A}} = \underline{\underline{B}}.\underline{\underline{C}}$ and $\underline{\underline{A}}$, $\underline{\underline{C}}$ are the 'direct' sums of row vectors $\underline{A}_i$, $\underline{C}_i$ then $\underline{A}_i = \sum_j B_{ij} \underline{C}_j$,

the sparsity of $\underline{\underline{B}}$ can be used but the vector length is the width of $\underline{\underline{C}}$. This will normally be ideal for the CRAY but too short for the CYBER 205. In the process of designing a good integral evaluation algorithm we took advantage of the fact that many blocks were to be evaluated and thus a parallel approach was suited. The same is true for the transformation (24), each step involves many $(n(n+1)/2)$ products so that we can perform all products in parallel. The algorithm in Fig. 7 must be modified so that the sort steps group together the same element from all matrices, e.g. $(\underline{\underline{G}}^{pq})_{\mu\nu}$ for all p,q that contain a μνth element. This is a better approach for the CYBER 205, and preferable up to n $\approx$ 80. A clear disadvantage is that IO activity is much greater except in the unlikely event that the full set of matrices at each step fits into main memory.

The initial sort in Fig. 7 is really an ordering since integrals are calculated in something approaching the correct order; the intermediate sort contributes $\approx 24.[[n]]$ operation times compared to $\approx 16n.[[n]]$ arithmetic operation times and thus is not usually significant in a full transformation.

A partial transformation is sufficient in MCSCF calculations

Fig. 7 The Full Transformation

steps	notes
order integrals to G^{pq}	G^{pq} is lower triangular with first [pq]-1 elements zero and contains elements $[pq\vert rs](1+\delta_{pq})^{-1}(1+\delta_{rs})^{-1}(1+\delta_{pq,rs})^{-1}$

(b)

$$H^{pq} = G^{pq} \cdot c$$

$$I^{pq} = \tilde{H}^{pq} \cdot c$$

$$J^{pq} = I^{pq} + \tilde{I}^{pq} \qquad J^{pq} \text{ is lower triangular}$$

sort $(J^{pq})_{ij}$ to $(J^{ij})_{pq}$

$$K^{ij} = J^{ij} c$$

$$L^{ij} = \tilde{K}^{ij} c$$

$$M^{ij} = L^{ij} + \tilde{L}^{ij} \qquad M^{ij} \text{ is lower triangular}$$

(a)

consider $(M^{ij})_{kl}$ as $M_{(ij)(kl)}$ supermatrix ordering $i \geq j$, $k \geq l$

$$G = M + \tilde{M} \qquad \text{transformed } G \text{ contains permutationally non-redundant elements}$$

and also in CI calculations on large molecules where a considerable number of electrons are to be frozen. In these cases we only
require the transformed integrals [pq|ij], [pi|qj] with i,j deno-
ting active orbitals and [pq|ii], [pi|qi] with i denoting frozen
orbitals, p,q can be considered as AO labels (incomplete transformation) or MO labels (with further restrictions on required integrals involving frozen orbitals). Fig. 8 shows an algorithm for
the incomplete partial transformation that it very similar to
one suggested by Werner and Meyer (15). Saunders and Van Lenthe
(21) have also given an algorithm but they choose to expand $\underline{\underline{G}}^{pq}$
to full lower triangular matrices (hence including permutationally
equivalent integrals) in order to better define the matrix structure. The objection to the algorithm in Fig. 8 is that more intermediate matrices must be available simultaneously so that the
algorithm degrades (slightly) when the basis size increases. The
objection to the algorithm of Saunders and Van Lenthe is that
a full sort must be performed initially, as well as intermediate
sorts (on smaller data structures).

Let us suppose that just one out-of-memory sort is to be
performed costing (optimistically) 24.[[n]] operation times. The
algorithms limit to 6m.[[n]] operation times (on the CYBER 205)
so that sorting is 20% even when m=16 (n is the basis size and
m the number of internal orbitals).

The contributions of MO integrals to the matrix-vector pro-
duct $\underline{\underline{H}}$ c = σ in direct CI has been analysed in detail by several
authors (21,25,26) and it has also been pointed out that the full
transformation can be avoided (21, 26-30). Let us denote orbitals
occupied in at least one reference configuration by i,j... (occu-
ied) and by a,b... those unoccupied (virtual), p,q... will denote
arbitrary orbitals. A partial transformation gives the generalised
coulomb and exchange matrices

$$(\underline{\underline{J}}^{ij})_{pq} = [pq|ij] \quad \text{and} \quad (\underline{\underline{K}}^{ij})_{pq} = [pi|qj] \qquad (26)$$

which includes all required MO integrals except [ab|cd] and
[ab|ci].

The contribution from these integrals can be written in
terms of an additional exchange matrix defined in the AO basis

$$(\underline{\underline{K}}^{P})_{\mu\nu} = \sum_{\lambda\sigma} G_{\mu\lambda\nu\sigma} \, v^{P}_{\lambda\sigma} \qquad (27)$$

where

$$(\underline{\underline{v}}^{P})_{\lambda\sigma} = \sum_{ab} \bar{C}_{Pab} \, c_{a\lambda} \, c_{b\sigma} \qquad (28)$$

and is a density-like matrix. $\bar{C}_{Pab}$ is the renormalised CI coeffi-

Fig. 8 Partial Transformation

step	notes
$V_i^p = \sum_q \underline{\underline{G}}^{pq} c_{iq}$	$\underline{\underline{G}}^{pq}$ lower triangular with $[pq]$ elements
$\underline{R}_i^{pq} = (\underline{\underline{G}}^{pq} + \tilde{\underline{\underline{G}}}^{pq}) \underline{c}_i$	$p \geq q$, p,q AOs, i,j MOs
$(\underline{\underline{T}}^{ij})_{pq} = \tilde{\underline{c}}_i \, \underline{R}_j^{pq}$	$i \geq j$
$\underline{R}_{\underline{\underline{Q}}_{ij}}^p = \sum_{q \geq p} c_{jq} \underline{R}_i^{pq} + \sum_{q < p} c_{jq} \underline{R}_i^{qp}$	$i \geq j$ pth row of $\underline{\underline{Q}}_{ij}$
$\underline{C}_{\underline{\underline{Q}}_{ij}}^p = \sum_{q \geq p} c_{iq} \underline{R}_j^{pq} + \sum_{q < p} c_{iq} \underline{R}_j^{qp}$	$i \geq j$ pth colomn of $\underline{\underline{Q}}_{ij}$
$\underline{\underline{Q}}_{ij} = \underline{R}_{\underline{\underline{Q}}_{ij}} + \underline{C}_{\underline{\underline{Q}}_{ij}}$	square matrix
$\underline{\underline{P}}_{ij} = \sum_q c_{iq} \underline{V}_j^q + \sum_q c_{jq} \underline{V}_i^q + \underline{\underline{T}}^{ij}$	$i \geq j$, lower triangular, diagonal elements are doubled.

cient of the state with internal identifier P, and external (virtual) MO labels a,b, the matrix $\underline{c}$ defines the MOs. Note that eq. (27) is an $n_p n^4$ process where n_p is the number of internal identifiers. A similar process is required if MO integrals are used, performing as $n_p n^e$ where n_e is the number of external MOs ($n_e < n$). Thus the initial saving from the partial instead of the full transformation (factor $<8n/m$) is repaid at each iteration (factor $(n_e/n)^4$). The balance depends on the partioning of n into m, n_e and deleted MOs and is rather delicate.

In direct direct-CI the motivation is to remove the necessity of explicitly processing long integral lists, and thus to replace IO by recalculation. AO integral-driven direct CI is already ripe for recalculation: the elements of $\underline{\underline{G}}$ in eq. (27) can be recalculated every iteration. Quite generally the coulomb and exchange matrices required can be written in the AO basis as

$$
\left.
\begin{aligned}
(\underline{\underline{J}}^{ij})_{\sigma\lambda} &= \sum_{\mu\nu} D_{\mu\nu}^{ij} \\[2mm]
(\underline{\underline{K}}^{ij})_{\nu\lambda} &= \sum_{\mu\sigma} D_{\mu\sigma}^{ij} \\[2mm]
(\underline{\underline{K}}^{P})_{\nu\lambda} &= \sum_{\mu\sigma} v_{\mu\sigma}^{P}
\end{aligned}
\right\}
\times G_{\mu\nu\sigma\lambda} \quad \text{with} \quad D_{\mu\nu}^{ij} = c_{i\mu}c_{j\nu}
$$

The similarity of these contractions with the construction of the closed shell Fock matrix in eq. (19) is obvious and there we were able to show that recalculation is quite feasible. In CI we should expect the same, at least for $\underline{\underline{K}}^{P}$. The iterative calculation of $\underline{\underline{J}}^{ij}$, $\underline{\underline{K}}^{ij}$ as a formal $m^2 n^4$ process competes with a single mn^4 partial transformation however the difference becomes smaller when frozen orbitals are taken into account (only $\underline{\underline{J}}^{11}$, $\underline{\underline{K}}^{11}$ required). Then we have an $(m_f + \frac{1}{2}m_a)n^4$ process competing agaisnt $(m_f + m_a)n^4$ where m_f, m_a are the number of frozen and active orbitals respectively and whenever m_f is large compared to m_a the difference is negligible.

That a large part of direct-CI can be written directly in terms of operator matrices in the AO basis can also be shown to be true for MCSCF. In this case the matrix-vector product is $\underline{\underline{E}} \underline{X} = -\underline{g}$ where $\underline{\underline{E}}$ is the Hessian, $\underline{g}$ the gradient and $\underline{X}$ the required solutions giving the orbital and CI coefficient rotations (in the exponential parameterization). Bacskay (31) has shown how this is possible in the special case of a single configuration. Consider i,j... to label orbitals doubly occupied in all reference states and t,u... those variably occupied, and a,b... the virtual orbitals. The contributions to the matrix-vector product from $\underline{\underline{J}}^{tu}$, $\underline{\underline{K}}^{tu}$ are sufficiently varied to warrant their explicit calculation and the CI-CI and CI-orbital parts of the Hession only consist of elements of these matrices (17). The terms in the orbital-orbital part requiring $\underline{\underline{J}}^{ij}$, $\underline{\underline{K}}^{ij}$ and $\underline{\underline{J}}^{it}$, $\underline{\underline{K}}^{it}$ must be expanded in the AO basis. For example the contributions to

$$\sum_j \sum_b E_{ia,jb} X_{jb} \quad \text{from} \quad \underline{\underline{J}}^{ij}, \underline{\underline{K}}^{ij}$$

can be written

$$\sum_{\mu\nu} Q_{\mu\nu} c_{a\mu} c_{i\nu}$$

where

$$Q_{\mu\nu} = \sum_{\lambda\sigma} P^{\mu\nu}_{\lambda\sigma} d_{\lambda\sigma} \quad \text{and} \quad d_{\lambda\sigma} = \sum_j \sum_b c_{j\lambda} c_{b\sigma} X_{jb}$$

The $\underline{\underline{P}}$ supermatrix is defined in eq. (20) and $\underline{\underline{Q}}$ is a Fock-like matrix. The contributions from $\underline{\underline{J}}^{iv}$, $\underline{\underline{K}}^{iv}$ are more complicated, contributions to

$$\sum_t \sum_b E_{ia,tb} X_{tb} \quad \text{and} \quad \sum_i \sum_a E_{ia,tb} X_{ia}$$

can be written

$$\sum_{\mu\nu} Q'_{\mu\nu} c_{a\mu} c_{i\nu} \quad \text{and} \quad \sum_v D_{tv}\left(\sum_{\mu\nu} Q_{\mu\nu} c_{b\mu} c_{v\nu}\right)$$

where

$$Q'_{\mu\nu} = \sum_{\lambda\sigma} P^{\mu\nu}_{\lambda\sigma} e_{\lambda\sigma} \quad \text{and} \quad e_{\lambda\sigma} = \sum_{tv} \sum_b D_{tb} X_{tb} c_{b\lambda} c_{v\sigma}$$

and $\underline{\underline{D}}$ is the first order density matrix of the CI wave-function in the active space.

Analogous manipulations can be carried out to eliminate the need to calculate explicitly MO integrals except those occuring in $\underline{\underline{J}}^{tu}$, $\underline{\underline{K}}^{tu}$. The calculations of $\underline{\underline{Q}}$ and $\underline{\underline{Q}}'$ are again contractions with the integrals ($\underline{\underline{P}}$) similar to the Fock matrix contraction leading us to believe that integral evaluation time will not be excessive in an AO integral driven scheme. However, it is also necessary to compare AO with MO integral driven schemes to compare absolute efficiencies since we have introduced an n^4 step in the iterative solution of the Newton-Raphson equations. The size of the Hessian increases rapidly with increasing problem size so that this in itself becomes problematic, an AO integral driven scheme avoids these problems. On the other hand it is not usually necessary to recompute the Hessian every MCSCF iteration, an

approximation being sufficient. For example, Siegbahn et al. (17)
show the convergence behaviour for HNO, the Hessian was calculated
in the first iteration (dimension 284) adjusted to remove negative
eigenvalues and inverted. The inverse was then used in the following five iterations when convergence was reached. The tranfor-
mation time is also reduced if the Hessian does not need to be
computed. In short an AO integral driven scheme is only likely
to be competitive when the Hessian is large and the full power
of quadratically convergent MCSCF is required for convergence.

Werner and Meyer (15) also show how the MCSCF equations can
be solved with only $\underline{\underline{J}}^{tu}$, $\underline{\underline{K}}^{tu}$ and $\underline{\underline{J}}^{11}$, $\underline{\underline{K}}^{11}$. Their method appriximates the Hessian where core orbitals $\overline{(i,j}....)$ are involved and
if these are energetically well separated the approximation is
good. Strict quadratic convergence is lost although this does
not appear to be severe. The coulomb and exchange matrices can
be calculated directly from AO integrals (an m^2n^4) process and
used immediately avoiding requiring them all simultaneously. Similar to the direct direct-CI case, if the ratio of core to valence (t,u....) orbitals is small then the m^2n^4 transformation
time approaches the mn^4 time.

DISCUSSION

The analysis that we have presented here is not sufficiently
detailed (except for the closed shell SCF case) to allow definitive
statements to made about the optimal use of vector processors.
It does seem clear however that iterative recalculation of AO
integrals with a knowledge of appropriate density matrices does
not need to be significantly slower than the more common single
calculation and iterative processing. The latter relies on fast
IO processing and a CP time dominated by floating point operations.
Vector processors exhibit (relatively) slow IO processing and,
unless especial care is taken in choice of algorithms, a CP time
dominated by trivial data movement. This shift in emphasis makes
recalculation (relatively) more attractive.

The storage of calculated integrals is in itself a problem,
one solved by iterative recalculation. Thus a wider range of molecular systems also comes within the scope of current computational
methods, e.g. large transition metal compounds (ferrocene) and
metal clusters (diffusion on surfaces). The direct SCF method
of Almlöf et al. (19) has been applied to complexes using more
than 500 primitive basis functions. The storage of all the integrals would obviously cause considerable problems.

We have only made passing reference to IO wait time since
it is a largely unpredictable aspect of modern computer systems.
Let us here suggest a data transfer rate to or from disc of 1

Mbyte per second (near the top end of technology) and a CP processing rate of 50 MFLOPS. One minute of CP time in an IO bound program would then require 6 hours of real time. The consequences are obvious.

REFERENCES

(1) Boys, S.F. 1950, Proc. R. Soc. London Ser A 200, pp. 542
(2) McMurchie, L.E. and Davidson, E.R. 1978, J. Comput. Phys. pp. 218
(3) Saunders, V.R., Science Research Council, Daresbury Laboratory, England, Publ.No. DL/SRF/P157
(4) Dupuis, M., Rys, J. and King, N.F. 1976, J. Chem. Phys. 65, pp. 111
(5) Hegarty, D. and Velde, G. van der 1983, Int. J. Quant. Chem. 23, pp. 1135
(6) Hegarty, D. and Velde, G.A. van der 1982, in Proceedings: "Int. Symp. on Vector Processing Applications", Colorado State University
(7) "ETO Multicenter Molecular Integrals", Eds., Weatherford, C.A. and Jones, H.W. 1982, D. Reidel Publishing Company, Holland
(8) Michels, H.H. in (7) pp. 103-121
(9) Raffenetti, R.C. 1973, J. Chem. Phys. 58, pp. 4452
(10) Ruedenberg, K., Schmidt, M., Gilbert, M. and Elbert, S.T. 1982, Chemical Physics 71, pp. 41
(11) Dacre, P.D. 1970, Chem. Phys. Lett. 7, pp. 47
(12) Elder, M. 1973, Int. J. Quant. Chem. 7, pp. 75
(13) Dupuis, M. and King, H.F. 1977, Int. J. Quant. Chem. 11, pp. 613
(14) Davidson, E.R. 1975, J. Chem. Phys. 62, pp. 400
(15) Werner, H.J. and Meyer, W. 1980, J. Chem. Phys. 73, pp. 2342
(16) Ruedenberg, K., Cheung, L.M. and Elbert, S.T. 1979, Int. J. Quant. Chem. 16, pp. 1069
(17) Siegbahn, P.E.M., Almlöf, L.M., Heiberg, A. and Roos, B.O. 1981, J. Chem. Phys. 74, pp. 2384
(18) Roos, B.O., Taylor, P.R. and Siegbahn, P.E.M. 1980, Chem. Phys. 48, pp. 157
(19) Almlöf, J., Korsell, K. and Faegri jr., K. 1982, J. Comput. Chem. 3, pp. 385
(20) Karlström, G. 1981, J. Comput. Chem. 2, pp. 33
(21) Saunders, V.R. and Lenthe, J.H. van 1983, Mol. Phys. 48, pp. 923
(22) Elbert, S.T. 1978, in "Numerical Algorithms in Chemsitry: Algebraic methods - LBL 8158", eds. Moler, C. and Shavitt, I. (Lawrence Berkeley Laboratory, Univ. of California, Berkeley)

(24) Ahlrichs, R., private communication
(25) Siegbahn, P.E.M. 1980, J. Chem. Phys. 72, pp. 1647
(26) Ahlrichs, R. 1981, in "Proceedings of the 5th European Semi-
     nar on Quantum Chemistry", eds. Duijnen, P.Th. van and
     Nieuwpoort, W.C., Univ. of Groningen, The Netherlands
(27) Zirz, C. and Ahlrichs, R. 1979, in "Electron Correlation:
     Proceedings of the Daresbury Study Weekend", eds. Guest,
     M.F. and Wilson, S., Daresbury Laboratory, England
(28) Meyer, W. 1976, J. Chem. Phys. 64, pp. 2901
(29) Dykstra, C.E., Schaefer, H.F. and Meyer, W. 1976, J. Chem.
     Phys. 65, pp. 2740
(30) Werner, H.J. and Reinsch, E.A. 1982, J. Chem. Phys. 76, pp.
     3144

MULTICONFIGURATION WAVEFUNCTIONS FOR MOLECULES: CURRENT APPROACHES

Thom. H. Dunning, Jr.

Theoretical Chemistry Group, Chemistry Division,
Argonne National Laboratory, Argonne, Illinois 60439

The generalized valence bond (GVB), fully optimized reaction space (FORS) and complete active space self-consistent field (CASSCF) methods are discussed. All of these approaches in their simplest form provide a multiconfiguration model for the electronic structure of molecules based on a well defined set of valence orbitals. As such they correct for a major deficiency in the Hartree-Fock model. While the GVB, FORS, and CASSCF methods permit the use of a larger orbital set, a scheme which provides a consistent description of molecular systems with an expanded set has yet to be developed.

I. INTRODUCTION

For many problems in chemistry single configuration wavefunctions, i.e., restricted Hartree-Fock (RHF) or unrestricted Hartree-Fock (UHF) wavefunctions, provide an excellent description of the electronic structure of the system. However, there are a number of problems, an example of which is the description of potential energy surfaces for the dissociation or reactions of molecules, for which multiconfiguration wavefunctions are required – a single configuration wavefunction is simply not able to properly describe the system over the range of nuclear coordinates of interest. Using multiconfiguration self-consistent field (MCSCF) methods, it is possible to construct wavefunctions that contain the important, geometry-dependent electron correlation effects in a lucid, compact form.

C. E. Dykstra (ed.),
Advanced Theories and Computational Approaches to the Electronic Structure of Molecules, 67–78.
© *1984 by D. Reidel Publishing Company.*

A critical problem in MCSCF calculations is in the choice of orbitals and configurations to be used in the wavefunction – the MCSCF method itself is only a mathematical technique for calculating wavefunctions, a chemical model based on a multi-configuration wavefunction requires a prescription for choosing the orbitals and configurations to be included in the wave-function. Further, unless due care is exercised in the choice of orbitals and configurations to be included in the calcula-tion, it is likely that the calculation will be biased and the results of questionable value. In recent years two general methods have been put forward for specifying the orbitals and configurations for use in MCSCF calculations: the generalized valence bond, GVB, method (1) and the fully optimized reaction space, FORS, method, (2). The complete active space self-consistent field, CASSCF, method (3) is conceptually related to the FORS method and in many instances is operationally equiva-lent to the more restricted FORS method; however, as we shall see below, a consistent scheme for extending the CASSCF wave-function beyond the FORS wavefunction has not yet been devel-oped. These methods have all been found to provide an adequate zero-order theoretical description of many molecular systems for a wide range of, but not all, applications.

Earlier attempts to develop multiconfiguration wavefunc-tions to describe the electronic structure of molecules include the optimized valence configuration, OVC, method of Das and Wahl (4) and the separated pairs method first suggested by Hurley, Lennard-Jones, & Pople (5) and developed in detail by Ruedenberg & coworkers (6).

II. THE GENERALIZED VALENCE BOND METHOD

The wavefunction in the generalized valence bond (GVB) method developed by Goddard & coworkers (1) is chosen to prop-erly describe dissociation to all asymptotic limits, including atomization; see also Lie & Clementi (7). The correlation effects associated with the incorrect behavior of the RHF wave-function upon dissociation are, of course, strongly geometry-dependent. There is, as the name implies, a close relationship to the valence bond (VB) method. As a simple example let us consider the results of GVB calculations on the hydrogen fluo-ride (HF molecule (8). The VB wavefunction for HF is (unnor-malized and omitting the core orbital)

$$\Psi_{VB}(HF) = \mathcal{A} \, 2s_F 2px_F 2py_F 2pz_F 1s_H \; \alpha\beta\alpha\beta\alpha\beta(\alpha\beta - \beta\alpha) \qquad (1)$$

In the VB method the orbitals are assumed not to change upon molecular formation. In the GVB wavefunction for HF, or in its

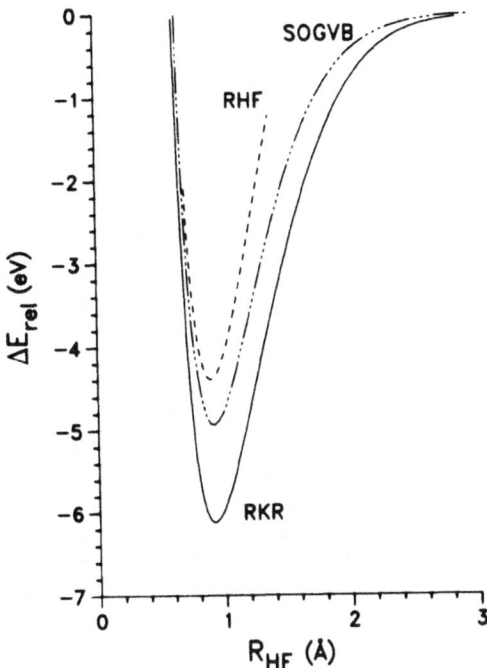

Figure 1. The RHF (9), GVB (8) and experimental (10) potential energy curves for the hydrogen fluoride molecule.

more tractable strongly orthogonal variant, SOGVB (9), the orbitals in Eq. (1) are optimized to allow for such changes. Thus, at $R_{HF} = \infty$, the SOGVB orbitals of HF are just the fluorine and hydrogen atomic orbitals, but at shorter separations the orbitals hydridize and delocalize so as to maximize the binding. The SOGVB method thus provides a theoretically consistent description of the HF molecule for all internuclear distances; see e.g., the calculated and experimental potential energy curves plotted in Figure 1.

The SOGVB wavefunction for HF can be equivalently written as a two configuration MCSCF wavefunction involving an orthogonal set of orbitals (again omitting the core orbital)

$$\Psi_{MCSCF}(HF) = c_1 \, \mathcal{A} \, 2\sigma^2 3\sigma^2 1\pi_x^2 1\pi_y^2 \quad \alpha\beta\alpha\beta\alpha\beta$$
$$- c_2 \, \mathcal{A} \, 2\sigma^2 4\sigma^2 1\pi_x^2 1\pi_y^2 \quad \alpha\beta\alpha\beta\alpha\beta \tag{2}$$

$$2\sigma_g^2 2\sigma_u^2 \begin{pmatrix} 3\sigma_g^2 \\ 3\sigma_g 3\sigma_u \\ 3\sigma_u^2 \end{pmatrix} \begin{pmatrix} 1\pi_{ux}^2 \\ 1\pi_{ux} 1\pi_{gx} \\ 1\pi_{gx}^2 \end{pmatrix} \begin{pmatrix} 1\pi_{uy}^2 \\ 1\pi_{uy} 1\pi_{gy} \\ 1\pi_{gy}^2 \end{pmatrix}$$

Figure 2. Direct product construction of the SOGVB wavefunction for the nitrogen molecule.

Variational optimization of the coefficients (c_1, c_2) and orbitals (2σ, 3σ, 4σ, $1\pi_x$, $1\pi_y$) in the MCSCF wavefunction Eq. (2) is equivalent to optimizing the orbitals in the SOGVB wavefunction Eq. (1). The MCSCF wavefunction, because it involves an orthogonal set of orbitals, provides a computationally efficient scheme for computing the SOGVB wavefunction; in fact, the SOGVB programs all have used a variant of this technique (12). The orbitals in Eq. (2) are referred to as the SOGVB natural orbitals.

Note that the first term in Eq. (2) is just the Hartree-Fock wavefunction for HF. At short internuclear distances $c_2 \ll c_1$ and the SOGVB and RHF wavefunctions differ little (although not negligibly; see Figure 1). At large internuclear distances, however, the energies of the two configurations become equal (degenerate) and $c_2 \to c_1$.

As a further example of the SOGVB method, let us consider the N_2 molecule. The configurations included in the SOGVB wavefunction for the N_2 molecule are obtained as a direct product of the three configurations for each of the pairs (the σ-bonding pair and the two π-bonding pairs); see Figure 2. The configurations generated in this way are listed in Table 1; there are ten (symmetry distinct) configurations in the SOGVB wavefunction of N_2. It should be noted that for some of the spatial configurations not all of the associated spinfunctions are included in the SOGVB wavefunction; these spinfunctions are not required to describe dissociation to two ground state nitrogen atoms.

The results of the SOGVB calculations on N_2 (13) are summarized in Table 2 where they are compared with the corresponding RHF calculations, full CI calculations with the SOGVB orbitals, recently reported CASSCF calculations (3a) and experiment. In Table 2 it is seen that the dissociation energy (D_e)

Table 1. Configurations in the SOGVB wavefunction for the nitrogen molecule. For brevity the core orbitals have been omitted.

Level[a]	$2\sigma_g$	$2\sigma_u$	$3\sigma_g$	$3\sigma_u$	$1\pi_{ux}$	$1\pi_{gx}$	$1\pi_{uy}$	$1\pi_{gy}$	Spin Coupling
0	2	2	2	0	2	0	2	0	
2	2	2	0	2	2	0	2	0	
b	2	2	2	0	0	2	2	0	
b	2	2	1	1	1	1	2	0	GF only
	2	2	2	0	1	1	1	1	GF only
4	2	2	0	2	0	2	2	0	
b	2	2	2	0	0	2	0	2	
	2	2	0	2	1	1	1	1	GF only
b	2	2	1	1	1	1	0	2	GF only
6	2	2	0	2	0	2	0	2	

[a]Excitation level relative to the RHF configuration.
[b]An additional configuration is obtained by replacing π_x by π_y.

Table 2. Summary of calculated (13,3a) and experimental (14) spectroscopic constants for the ground state of the nitrogen molecule. Units are as indicated.

	RHF	SOGVB	SOGVB-CI	CASSCF	Expt'l
R_e (Å)	1.067	1.097	1.106	1.108	1.0977
D_e (eV)	5.08	7.21	8.93[a]	8.76[a]	9.905
ω_e (cm^{-1})	2757.	2374.	2330.	2332.	2358.027
B_e	2.115	2.000	1.970	1.962	1.9980

[a]Using a much larger basis set and including all of the valence orbitals in a FORS calculation, Ruedenberg et al. (2d) obtained a dissociation energy of 9.06 eV.

obtained from the SOGVB calculations is more than 2 eV larger than that obtained from the RHF calculations, thus correcting a serious deficiency in the RHF description of N_2. Carrying out a full CI calculation using the SOGVB orbitals (176 configurations in D_{2h} symmetry) increases D_e by another 1.7 eV, resulting in a dissociation energy of 8.93 eV; for comparison, the measured D_e is 9.905 eV. Clearly, the SOGVB wavefunction

includes many of the terms of importance in describing the
electron correlation effects associated with molecular formation, while other important effects can be taken into account
by including additional configurations involving the SOGVB
orbitals.

In summary, then, the advantages of the SOGVB method
include:

(1) Configurations can be chosen based on simple VB concepts to include those correlations effects likely to be most
geometry-dependent.

(2) The SOGVB form of the wavefunction yields a readily
interpretable model of the electronic structure of the system.

(3) The calculations are sufficiently tractable that they
can be applied to relatively large systems.

However, not all important correlation effects can be accounted
for with the simple SOGVB wavefunction, see, e.g., N_2 above.
For example, for some molecules more than one VB structure is
possible; to properly describe these molecules requires a projected SOGVB wavefunction (15).

III. THE FULLY OPTIMIZED REACTION SPACE AND COMPLETE ACTIVE SPACE SELF-CONSISTENT-FIELD METHODS

The fully optimized reaction space (FORS) and complete
active space self-consistent-field (CASSCF) methods emphasize
the choice of the orbitals to be included in the calculation,
rather than the choice of configurations and orbitals as in the
SOGVB method. These methods divide the orbitals into three
sets: inactive, active and virtual. The configurations in the
FORS/CASSCF wavefunction then include all configurations which
can be constructed with the inactive orbitals doubly occupied
and the virtual orbitals unoccupied, i.e., all possible configurations (full CI) are constructed within the active space. The
FORS method was proposed by Ruedenberg & coworkers (2) and the
CASSCF method was put forward by Siegbahn, Roos, & coworkers
(3).

The main advantage of the above methods is that once the
partitioning between the active, inactive, and virtual orbitals
is established, the wavefunction is completely specified. The
main disadvantage of these methods is that the number of configurations which must be included in the calculation increases
dramatically with the number of electrons in the active space.
This is clearly illustrated in Figure 3 which plots the number

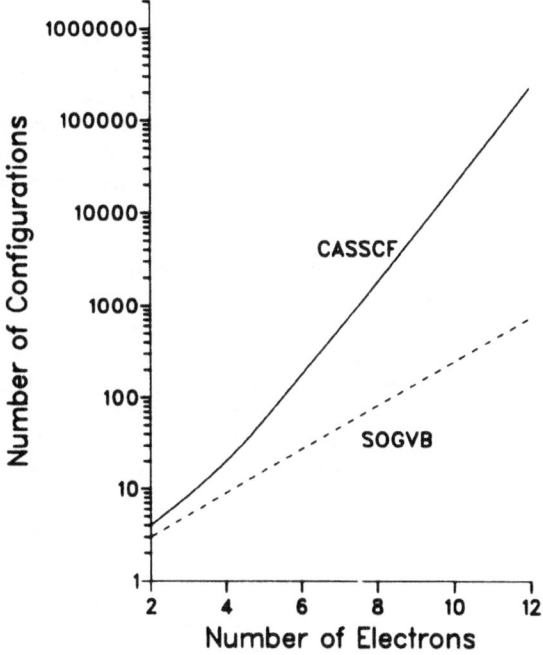

Figure 3. Number of configurations in the SOGVB and CASSCF wavefunctions for an n electron singlet state. There are n orbitals in the active space and no symmetry is assumed.

of configurations included in the SOGVB and FORS/CASSCF wavefunction for a singlet state with n electrons in n orbitals and no symmetry; for the twelve electron case there are only 729 configurations in the SOGVB wavefunction compared to 226,512 configurations in the FORS/CASSCF wavefunction.

The FORS method and one variant of the CASSCF method are based on an active orbital set whose composition, if not form, is identical to that used in SOGVB calculations. This set is referred to as the valence orbital set and is composed of all molecular orbitals which correlate with the valence shell atomic orbitals, i.e., the 1s orbital in hydrogen, the ($2s$, $2p_x$, $2p_y$, $2p_z$) orbitals in the first row atoms (Li–Ne), etc. In most cases the results of the FORS or CASSCF calculations are found to be quite similar to those obtained from the corresponding full CI calculations in the SOGVB orbital space.

This point is clearly illustrated in Table 2 where it is seen that the results of the full SOGVB-CI (13) and CASSCF (3a) calculations on N_2 differ little. In fact, the slight variations in the basis sets and orbital sets used in the two calculations could well explain much of the observed differences; the basis set for the CASSCF calculations included only a single Gaussian 3d function on each nitrogen while the SOGVB-CI calculation included two 3d Gaussian functions on each nitrogen contracted to one function and, in addition, the CASSCF calculations have one fewer orbital in the active space than the full SOGVB-CI calculations.

A larger active orbital set is often used in CASSCF calculations. However, as discussed in the following section, it is possible to introduce serious biases into the calculation if the additional orbitals are improperly chosen. Although arguments have been advanced for including the orbitals necessary to describe the united atom limit in CASSCF calculations (3), at this time it is not clear whether the resulting set provides a consistent means of expanding the valence orbital set.

In summary, the advantages of the FORS/CASSCF method include:

(1) The wavefunction is completely specified by the partitioning of the orbital set into inactive, active and virtual orbitals.

(2) The method is more general than the SOGVB method, allowing for the inclusion of correlation effects beyond those associated with improper dissociation.

The disadvantages of the FORS/CASSCF method include:

(1) The number of configurations included in the wavefunction is a strong function of the number of electrons in the active space.

(2) Biases may arise if the orbital set is expanded beyond the simple valence orbital set.

(3) Erratic changes may occur in the orbitals included in the calculation.

The last point above needs to be discussed further. Unlike the SOGVB method, the FORS/CASSCF method allows considerable flexibility in the orbitals included in the calculation. This can give rise to problems as the orbital set may change character from one geometry to the next, each set, of course, leading to the lowest possible energy at the given geometry. Such changes, if they occur, will lead to erratic behavior in the FORS/CASSCF energies, properties, etc. Further, if different functional forms of the orbitals lead to nearly indistinguishable total wavefunctions, the MCSCF calculation may either converge poorly or converge to local minima rather than the global minimum. This problem is expected to be less severe for a FORS wavefunction or a CASSCF wavefunction based on the valence orbital set as these wavefunctions tend to be dominated by the SOGVB configurations.

IV. BEYOND MULTICONFIGURATION SELF-CONSISTENT FIELD WAVEFUNCTIONS BASED ON THE VALENCE ORBITAL SET

As noted above the valence orbital set forms the basis of the SOGVB and FORS methods as well as one common form of the CASSCF method. This appears to be a judicious choice for extensions of this set can lead to serious inconsistencies. As an example let us consider the hydrogen fluoride molecule again (8). A more accurate wavefunction for HF would allocate a separate orbital for each electron, i.e., (with the doubly occupied core orbital omitted)

$$\mathcal{A} \, 2\sigma_a 2\sigma_b 3\sigma_a 3\sigma_b 1\pi_{xa} 1\pi_{xb} 1\pi_{ya} 1\pi_{yb} \, \lambda_{spin} \tag{3}$$

where λ_{spin} is the appropriate spinfunction. For the separated atoms this corresponds to adding a 3s, $3p_x$, and $3p_y$ orbital to the valence orbital set. In the molecule 5σ, $2\pi_x$, and $2\pi_y$ orbitals are added to the set given in Eq. (2). Finally, in the united atom the orbitals in Eq. (3) provide a complete set of n = 3 orbitals (3s, $3p_x$, $3p_y$, $3p_z$). It is for this reason that arguments have been made (3) to include the 2π-orbitals in the CASSCF orbital set; in the united atom limit the SOGVB wavefunction adds only one additional orbital, a $3p_z$ orbital, to the RHF set, thus, providing an unbalanced description of the united atom.

Calculations with the above wavefunction, Eq. (3), and a polarized double zeta basis set and with λ_{spin} being the perfect-pairing spin coupling yields a D_e of 4.84 eV in line with that obtained from the corresponding SOGVB calculations, namely, 4.74 eV. However, if the natural orbitals from Eq. (3) are used in a CI calculation, the predicted D_e is 5.65 eV, nearly 0.9 eV larger than that obtained from the SOGVB-CI calculations, 4.77 eV. Further, the estimated D_e at the CI limit for this basis set is just 5.7 ± 0.1 eV. Thus, there is not only a large difference between the CI calculations with the SOGVB orbitals and the orbitals from Eq. (3), but, further, the CI calculations based on Eq. (3) yield a D_e within 0.05 ± 0.1 eV of the exact result (for the basis set used).

To put these results into perspective note that the CI wavefunction based on the orbitals from Eq. (3) includes the double excitation

$$(3\sigma 1\pi) \;\rightarrow\; (4\sigma 2\pi)$$

where $(3\sigma, 4\sigma)$ are the bonding and antibonding σ natural orbitals and $(1\pi, 2\pi)$ are the two π natural orbitals. Near R_e this configuration accounts for interpair correlation effects involving the σ bond pair and the fluorine π lone pairs [at the united atom limit the above excitation corresponds to $(2p_z 2p_x)$ $\rightarrow$ $(3p_z 3p_x)$]. At large internuclear distances, however, the coefficient for this configuration vanishes. That is, at the separated atom limit the CI wavefunction based on the orbital set from Eq. (3) does not include any interpair terms to correlate the electrons in the singly occupied $2p_z$ orbital and the electrons in the doubly occupied $2p_x$ and $2p_y$ orbitals. The $3p_z$ orbital needed to describe this effect is not present in the orbital set based on Eq. (3). Thus, the CI wavefunction based on Eq. (3) is biased against the separated atom limit and the calculated dissociation energy is far too large.

As this example illustrates, going beyond MCSCF models based on the simple valence orbital set requires considerable care. Any models proposed must be carefully checked for both theoretical consistency (as in the above) and computational stability (see, e.g., the remarks at the end of Section III).

ACKNOWLEDGMENT

Work performed under the auspices of the Office of Basic Energy Sciences, Division of Chemical Sciences, U. S. Department of Energy, under Contract W-31-109-Eng-38.

REFERENCES

1. Goddard, III, W. A., 1967, Phys. Rev. 157, pp. 81–93; Ladner, R. C. and Goddard, III, W. A., 1969, J. Chem. Phys. 51, pp. 1073–1087; Hunt, W. J., Hay, P. J. and Goddard, III, W. A., 1972, J. Chem. Phys. 57, 738–748.

2. Ruedenberg, K. and Sundberg, K. R., 1976, in Quantum Science, ed. Calais, J-L., Goscinski, O., Linderberg, J. and Ohrn, Y., Plenum Publishing Corporation, New York, pp. 505–515; Ruedenberg, K., Schmidt, M. W., Gilbert, M. M. and Elbert, S. L., 1982, Chem. Phys. 71, pp. 41–49; Ruedenberg, K., Schmidt, M. W. and Gilbert, M. M., 1982, Chem. Phys. 71, pp. 51–64; Ruedenberg, K., Schmidt, M. W.,

Gilbert, M. M. and Elbert, S. T., 1982, Chem. Phys. 71, pp. 65–78.

3. Roos, B. O., Taylor P. R. and Siegbahn, P. E. M., 1980, Chem. Phys. 48, pp. 157–173; Roos, B. O., 1980, Intern. J. Quantum Chem.: Quantum Chemistry Symposium 14, pp. 175–189.

4. See, e.g., the work summarized in Wahl, A. C. and Das, G., 1977, "The Multiconfiguration Self-consistent Field Method," in Methods in Electronic Structure Theory, ed. Schaefer, III, H. F., Plenum Press, New York, chap. 3.

5. Hurley, A. C., Lennard-Jones, J. and Pople, J. A., 1953, Proc. Roy. Soc. (London) A220, pp. 446–455.

6. Miller, K. J. and Ruedenberg, K., 1968, J. Chem. Phys. 48, pp. 3414–3443; Silver, D. M., Mehler, E. L. and Ruedenberg, K., 1970, J. Chem. Phys. 52, pp. 1174–1180; Mehler, E. L., Ruedenberg, K. and Silver, D. M., 1970, J. Chem. Phys. 52, pp. 1181–1205.

7. Lie, G. C. and Clementi, E., 1974, J. Chem. Phys. 60, pp. 1275–1288; Lie, G. C. and Clementi, E., 1974, J. Chem. Phys. 60, pp. 1288–1296.

8. Dunning, Jr., T. H., unpublished; see also Dunning, Jr., T. H., 1976, J. Chem. Phys. 65, pp. 3854–3862.

9. Cade, P. E. and Huo, W. M., 1967, J. Chem. Phys. 47, pp. 614–648.

10. DiLonardo, G. and Douglas, A. E., 1973, Can. J. Phys. 51, pp. 434–445.

11. Bobrowicz, F. W., 1974, Ph.D. Thesis, California Institute of Technology, Pasadena, California; Moss, B. J., Bobrowicz, F. W. and Goddard, III, W. A., 1975, J. Chem. Phys. 63, pp. 4632–4639.

12. Bobrowicz, F. W. and Goddard, III, W. A., 1977, "The Self-Consistent Field Equations for Generalized Valence Bond and Open-Shell Hartree-Fock Wave Functions," in Methods in Electronic Structure Theory, ed. Schaefer, III, H. F., Plenum Press, New York, chap. 4.

13. Dunning, Jr., T. H., Cartwright, D. C., Hunt, W. J., Hay, P. J. and Bobrowicz, F. W., 1976, J. Chem. Phys. 64, pp. 4755–4766.

14. Rosen, B., 1970, <u>Spectroscopic Data Relative to Diatomic Molecules</u>, Pergamon, New York.

15. Levin, G. and Goddard, III, W. A., 1975, J. Am. Chem. Soc. 97, pp. 1649-1656; Voter, A. F. and Goddard, III, W. A., 1981, Chem. Phys. 57, pp. 253-259.

INTERNALLY CONTRACTED MCSCF-SCEP CALCULATIONS

Hans-Joachim Werner* and Ernst-Albrecht Reinsch

Fachbereich Chemie, Universitaet Frankfurt
D-6000 Frankfurt/M, West Germany

Abstract

The internally contracted MCSCF-SCEP method is reviewed. Some technical details concerning the optimal organization and vectorization of the program are discussed, and timings for calculations on a CRAY-1 with up to 637524 configurations and up to 172908 variational parameters are analyzed. Some examples of internally contracted and uncontracted calculations are compared. The stability of calculated electric dipole moments and electronic transition moments with respect to the number of reference configurations and internal orbitals is investigated. Finally, a brief survey of recent applications of this method for calculating potential energy functions and radiative lifetimes of electronically excited states is presented.

I.   INTRODUCTION

In recent years considerable progress has been made in the calculation of accurate multireference configuration expansions (MR-CI). Conventional MR-CI procedures have been applied for quite a long time, (1-4,6) but these methods could handle only a limited number of configurations ($\approx 2 \cdot 10^4$), and therefore configuration selections (2) or transformations to pseudo natural

*present address:

Los Alamos National Laboratory, University of California, Mail Stop J579, Los Alamos, New Mexico 87545, USA

79

C. E. Dykstra (ed.),
Advanced Theories and Computational Approaches to the Electronic Structure of Molecules, 79–105.
© *1984 by D. Reidel Publishing Company.*

orbitals (5,6) were mostly required. The goal of including in the wavefunction all singly and doubly excited configurations relative to an arbitrary set of reference configurations, and thus avoiding the configuration selection and the related uncertainties of the results, has been achieved by the development of direct multireference CI procedures (7-12). In these methods the required eigenvectors are calculated iteratively without storing the Hamiltonian matrix, which makes possible the optimization of much longer configuration expansions than with conventional CI methods.

A severe disadvantage of most direct MR-CI methods is the fact that the number of configurations and the computational effort strongly depends on the number of reference configurations (13). Already for simple diatomic molecules which contain two first row atoms, sometimes 20-30 reference configurations are necessary in order to describe dissociation properly. When using large basis sets as required for calculating accurate molecular properties, one can then generate typically 10^5-10^6 singly and doubly substituted configuration state functions. Although configuration expansions of this size can now be handled even on minicomputers (11), such calculations are still quite expensive.

The number of configurations can be reduced without losing much accuracy if only the "first-order interacting configuration space" (14-16) is considered, i.e. by omitting all configurations which do not interact directly with the reference wave function. This concept can only partly be exploited with schemes which depend on particular spin couplings, as is the case for methods based on the graphical unitary group approach (GUGA) (7,8,11), for example. Meyer has shown that a set of configurations which exactly span the first order interacting space can easily be generated by successively applying spin-coupled pair annihilation and creation operators to the reference configurations (16). This method is applied in the MCSCF-SCEP method (9) which is reviewed in the present paper.

The number of variational parameters and hence the computational effort can be further reduced by contracting certain classes of configurations with fixed coefficients. So far, two contraction schemes have been proposed, both of which make use of the fact that the configuration state functions can be factorized into an internal and an external part. The first scheme, which has been proposed and applied by Siegbahn (17), is called the "external contraction". In this case all configurations Ψ_P^{ab} and Ψ_S^a which belong to the same internal hole states P or S but have different external orbitals a,b are contracted with coefficients $(C_P)_{ab}$ and $(c_S)_a$, respectively, which are obtained by first order perturbation theory. Then, for each internal state only one parameter is optimized by performing a relatively small CI calculation. The computational effort for this calculation is somewhat larger than that of one direct CI iteration for the uncontracted wave function, i.e., the total cost is reduced by a factor of 5-10, but it still depends on the number of reference configurations. It has been found that the external contraction degrades the correlation energy typically by 2-3 per cent. However, the errors of non-energetic properties, such as dipole moments, can be much larger.

A third possibility to reduce the cost of an MR-CI calculation is to apply the "internal contraction" scheme. It was first proposed by Meyer (16), and has been used for the first time in the MCSCF-SCEP method of the present authors (9). In this case the contracted configurations are generated by applying pair annihilation and creation operators to the complete multiconfiguration reference wave function. Thus, those configurations which are generated by applying the same operators to different reference configurations are contracted with the expansion coefficients of the reference configurations. Provided that the reference wave function has been optimized by an MCSCF procedure and a second state is not extremely close, the internal contraction has been found to be an excellent approximation. In fact, this is to be expected since in terms of perturbation theory the contracted and uncontracted wave functions differ only in second order. As will be shown in the last section, the loss of correlation energy is typically only 0.1-0.3 per cent, and also the changes of other properties, such as dipole or transition moments, are negligibly small. The computational effort is reduced approximately by the same factor as the number of variational parameters. The latter number is always smaller than a certain threshold, which depends only on the number of correlated internal orbitals and not on the number of reference configurations. Hence, the above-mentioned main drawback of the MR-CI method, namely the strong dependence of the cost on the number of reference configurations, has been eliminated to a large extent. The price one has to pay for the reduction of the variational parameters is a more complicated structure of the internal hole states. As will be shown in the following sections, the coupling coefficients needed for calculating $\underline{H} \cdot \underline{c}$ are simple overlap or transition density matrix elements between a set of internal "core functions". Since the contracted core functions are linear combinations of the uncontracted ones, about the same number of coupling coefficients is obtained in both cases. That the internal contraction nevertheless leads to a reduction of the computation time is due to the following facts: a) the number of external exchange operators calculated in each iteration is reduced, and b)

the number of matrix multiplications is considerably reduced.
Instead,only simple contractions of two-electron integrals have to
be performed.

A very important step in the development of efficient direct CI
procedures was made by recognizing that the coupling coefficients
can always be factorized into an internal and an external part
(7,16,18). In the theory of self-consistent electron pairs (SCEP)
formulated in 1976 by Meyer (18) for the case of closed-shell
one-determinant reference functions, it was shown that the external
coupling coefficicients can be completely removed by a particular
renormalization of the doubly external configurations. In 1981 the
present authors generalized the SCEP method for general internally
contracted or uncontracted reference wave functions (9). As is
characteristic of the SCEP method, no external coupling
coefficients appear in the formalism, and the CI coefficients as
well as the two-electron integrals with at most two external
orbitals are grouped into vectors or matrices. In this way any
logic can be removed from the innermost loops, and the vector $\underline{H} \cdot \underline{c}$
is obtained in terms of simple matrix operations. Furthermore, no
full transformation of the two-electron integrals into the MO basis
is required, which is advantageous if large basis sets are
employed. Due to its matrix structure, the SCEP method is equally
well suited for mini and for super computers. In fact, our program
was developed on a slow scalar machine, but after some minor
changes (such as expanding triangular matrices to square form) a
high efficiency on a CRAY-1 computer was achieved. It should be
noted that at the same time the first results obtained with our
internally contracted MCSCF-SCEP method were published, Ahlrichs
(19) presented a very similar formalism for the uncontracted MR-CI
case. Recently, Saunders and van Lenthe (12)developed an
uncontracted MR-CI program which also makes use of SCEP-like
techniques.

In the present paper, the theory of the internally contracted
MCSCF-SCEP method will be reviewed first. Some details of the
implementation of our method on a vector computer will then be
discussed, and some typical timings will be presented. Furthermore,
some examples of internally contracted and uncontracted calculations will be compared. Finally, a brief survey of recent
applications of our method for calculating accurate dipole and
transition moment functions will be given.

II. DEFINITION OF THE CONFIGURATION SPACE

The MCSCF-SCEP wave functions are constructed from a set of
orthonormal orbitals which is divided into three parts: core

orbitals, which are doubly occupied in all configurations and not correlated; valence orbitals, which are occupied in any of the reference configurations; and external orbitals, into which one or two valence electrons may be excited. In the following the valence orbitals will be denoted by the labels $i,j...$, the external orbitals by the labels $a,b...$, and either or both by the labels $r,s...$ Summations are over the corresponding subspaces only.

The N-electron wavefunction is approximated by the following expansion:

$$\Psi = \sum_I c_I \Psi_I + \sum_S \sum_a (\underline{c}_S)_a \psi_S^a + \sum_P \sum_{ab} (\underline{C}_P)_{ab} \psi_P^{ab} \quad , \qquad (1)$$

where the configurations Ψ_I, ψ_S^a, and ψ_P^{ab} have zero, one, and two electrons, respectively, in the external subspace. The indices S and P denote N-1 and N-2 electron hole states, which will be defined below. As indicated above, the expansion coefficients are grouped into vectors $\underline{c}_S$ and matrices $\underline{C}_P$. The matrices $\underline{C}_P$ are symmetric for singlet pairs and antisymmetric for triplet pairs, i.e. $(\underline{C}_P)_{ab} = p(\underline{C}_P)_{ba}$, with $p=1-2S$. For an uncontracted MR-CI wave function the configurations are generated by applying successively spin-coupled pair annihilation and creation operators to the reference configurations ψ_K (16,9):

$$\psi_{ijp,K}^{rs} = N \sum_m B_{rs,pm}^+ B_{ij,pm} \psi_K \quad , \qquad (2)$$

where N is a normalization factor, and

$$B_{ij,pm} = \begin{cases} n_i^\alpha n_j^\alpha & (p=-1, \ m=1) \\ 2^{-1/2}(n_i^\alpha n_j^\beta - p \ n_i^\beta n_j^\alpha) & (p=\pm 1, \ m=0) \\ n_i^\beta n_j^\beta & (p=-1, \ m=-1) \ . \end{cases} \qquad (3)$$

The configurations $\psi_{ijp,K}^{rs}$ are eigenfunctions of the spin operators S^2 and S_z, provided the same is true for the ψ_K. Similarly, the internally contracted configurations are generated by applying the operators to the whole reference function Ψ_0:

$$\psi_{ijp}^{rs} = \sum_m B_{rs,pm}^+ B_{ij,pm} \Psi_0 = \sum_K a_K \psi_{ijp,K}^{rs} \quad , \qquad (4)$$

with

$$\Psi_0 = \sum_K a_K \psi_K \quad . \qquad (5)$$

The configurations Ψ_{ijp}^{rs} exactly span the first order interacting space of Ψ_0 (16). By factoring out the external creation operators, the doubly external configurations can be written as

$$\Psi_P^{ab} = 2^{-1/2} \sum_m B_{ab,pm}^+ \, \Phi_{Pm} \quad , \tag{6}$$

where Φ_{Pm} are N-2 electron core functions. The index P specifies the internal annihilation operators and their spin coupling, i.e. P=(ijp) for contracted wavefunctions, and P=(ijp,K) for the uncontracted ones. In the following formalism no particular information about the structure of the core functions is required, and no distinction between the two cases is necessary. For convenience in later expressions, the normalization factor N has been incorporated into the core functions and is chosen such that

$$\sum_m \langle \Phi_{Pm} | \Phi_{Pm} \rangle = 1 \quad . \tag{7}$$

Hence, the doubly external configurations are normalized as

$$\langle \Psi_P^{ab} | \Psi_P^{ab} \rangle = 1/(2-\delta_{ab}). \tag{8}$$

By means of this particular normalization, the "external coupling coefficients" are removed from the formalism. Similarly, the singly external configurations may be written as

$$\Psi_S^a = \sum_\sigma \eta_a^\sigma \Phi_{S\sigma} \qquad (\sigma=\alpha,\beta) \quad , \tag{9}$$

where $\Phi_{S\sigma}$ are N-1 electron core functions, which are normalized as:

$$\sum_\sigma \langle \Phi_{S\sigma} | \Phi_{S\sigma} \rangle = 1 \tag{10}$$

(for details see Ref. 9). The internal configurations Ψ_I are generated as in Eq. 4, with r,s internal, and then normalized.

In general, the function sets $\{\Psi_I\}$, $\{\Psi_S^a\}$, and $\{\Psi_P^{ab}\}$ defined above are neither orthogonal nor linearly independent. Linearly independent subsets can be obtained by diagonalizing the strongly blocked overlap matrices

$$\sigma_{IJ} = \langle \Psi_I | \Psi_J \rangle \quad , \tag{11}$$

$$\sigma_{ST} = \Sigma \ \langle \Phi_{S\sigma} | \Phi_{T\sigma} \rangle \ , \tag{12}$$

$$\sigma_{PQ} = \overset{\sigma}{\underset{m}{\Sigma}} \ \langle \Phi_{Pm} | \Phi_{Qm} \rangle \ , \tag{13}$$

and eliminating those core functions which have the largest coefficients in the eigenvectors corresponding to small (e.g. $<10^{-3}$) eigenvalues. The remaining core functions may then be subject to either the canonical or the symmetrical orthonormalization scheme. However, as is shown in section III, it is advantageous for the internal and singly external configurations to perform the transformations at a later stage of the calculation.

III. THE ITERATIVE SCEP PROCEDURE

In our configuration basis the Schroedinger equation takes the form

$$\langle \Psi_I | H-E | \Psi \rangle = 0 \quad \text{for all } I \ , \tag{14}$$

$$\langle \Psi_S^a | H-E | \Psi \rangle = 0 \quad \text{for all } S,a \ , \tag{15}$$

$$\langle \Psi_P^{ab} | H-E | \Psi \rangle = 0 \quad \text{for all } P,a,b \ . \tag{16}$$

In order to solve these equations we apply the following iteration scheme: the coefficients of the singly and doubly external configurations are updated using first order perturbation theory

$$\Delta(\underline{c}_S)_a = - \langle \Psi_S^a | H-E | \Psi \rangle / \langle \Psi_S^a | H-E | \Psi_S^a \rangle \ , \tag{17}$$

$$\Delta(\underline{C}_P)_{ab} = - \langle \Psi_P^{ab} | H^-E | \Psi \rangle / E_P^{ab} \ , \tag{18}$$

with

$$E_P^{ab} \simeq \langle \Psi_P^{ab} | H-E | \Psi_P^{ab} \rangle / \langle \Psi_P^{ab} | \Psi_P^{ab} \rangle \ . \tag{19}$$

Here Ψ is the wave function of the previous iteration and $E=E_0+E_{corr}$ is the corresponding total energy. Instead of E_{corr} other values can be used in order to obtain CEPA type (5) solutions. For instance, with $E_{corr}=0$ the linearized coupled cluster equations are obtained. The external configurations are then contracted as

$$\Psi_{ext} = \underset{S \ a}{\Sigma \ \Sigma} \ (\underline{c}_S)_a \ \Psi_S^a + \underset{P \ ab}{\Sigma \ \Sigma} \ (\underline{C}_P)_{ab} \Psi_P^{ab} \ , \tag{20}$$

and the coefficients of the internal configurations are optimized by performing a small conventional CI for the contracted wave function:

$$\Psi = \sum_I c_I \Psi_I + \alpha_{ext}\Psi_{ext} \ . \tag{21}$$

A new energy eigenvalue is obtained simultaneously. This method is easily generalized to a multistate optimimization procedure by adding to Eq.(21) the external functions for the other states. Furthermore, convergence can be improved by expanding Ψ_{ext} into the increments obtained in each iteration as in the Davidson procedure (20), but this has not been done in our present implementation.

In order to obtain the matrix elements needed for the above procedure in a compact form, we define the generalized Coulomb and exchange matrices

$$(\underline{J}^{kl})_{ab} = (kl|ab) \ , \tag{22}$$

$$(\underline{K}^{kl})_{ab} = (ak|lb) \ , \tag{23}$$

and the integral vectors

$$(\underline{I}^{jkl})_a = (kl|ja) \ . \tag{24}$$

Furthermore, we define external exchange operators

$$[\underline{K}(\underline{C}_P)]_{rs} = \sum_{ab} (\underline{C}_P)_{ab}(ra|bs). \tag{25}$$

These operators are the only quantities which depend on the two-electron integrals over four external orbitals. They can be calculated directly from the two-electron integrals in the AO basis as

$$[\underline{K}(\underline{C}_P)]_{rs} = \sum_{\mu\nu} \left\{ r_\mu s_\nu \left[\sum_{\substack{\rho\sigma \\ ab}} (\underline{C}_P)_{ab} \, a_\rho b_\sigma \right] (\mu\rho|\nu\sigma) \right\} \ , \tag{26}$$

where $\underline{r}$, $\underline{s}$, $\underline{a}$, $\underline{b}$ are the orbital vectors. Hence, a full transformation of the two-electron integrals into MO basis is not required. Finally, we define a core Fock operator

$$\underline{F} = \underline{h} + \sum_c (2\underline{J}^{cc} - \underline{K}^{cc}) \ , \tag{27}$$

where the index c runs over the core orbitals only. By means of this operator the core orbitals are removed from the remaining formalism. The required matrix elements are now obtained in terms of matrices $\underline{G}_P$ and vectors $\underline{g}_S$:

$$\langle \Psi_P^{ab} | H-E | \Psi \rangle = \left[1/2(\underline{G}_P + p\ \underline{G}_P) - E\ \underline{C}_P \right]_{ab} \ , \tag{28}$$

$$\langle \Psi_S^{a} | H-E | \Psi \rangle = \left[\underline{g}_S - E\ \underline{c}_S \right]_{a} \ , \tag{29}$$

$$\langle \Psi_I | H-E | \Psi_{ext} \rangle = g_I \ , \tag{30}$$

$$\langle \Psi_{ext} | H | \Psi_{ext} \rangle = \sum_S \underline{c}_S^+ \cdot \underline{g}_S + \sum_P tr(\underline{C}_P^+ \cdot \underline{G}_P) - \sum_I g_I c_I \ , \tag{31}$$

$$\langle \Psi_{ext} | \Psi_{ext} \rangle = \sum_S \underline{c}_S^+ \cdot \underline{c}_S + \sum_P tr(\underline{C}_P^+ \cdot \underline{C}_P) \ , \tag{32}$$

The matrix elements $\langle \Psi_I | H | \Psi_J \rangle$ are calculated explicitly using standard methods. The matrices $\underline{G}_P$, the vectors $\underline{g}_S$, and the elements g_I are given by the following matrix expressions:

$$\underline{G}_P = \underline{K}(\underline{D}_P) + 2^{-1/2} \sum_{kl} \underline{K}^{kl} \sum_I c_I\ \sigma(P,klI)$$

$$+ \sum_Q \{ \sum_{kl} [\underline{J}^{kl} \alpha_{lk}(P,Q) - \underline{K}^{kl} \beta_{lk}(P,Q)] + \underline{1}\ \gamma(P,Q) + 2\ \delta_{PQ} \underline{F} \} \cdot \underline{C}_Q$$

$$+ 2^{1/2} \sum_S \{ \sum_j \underline{F}^j \sigma(P,jS) + \sum_{jkl} [\ \underline{I}^{jkl} \alpha_{lk}(P,jS)] \} \cdot \underline{c}_S^+ \tag{33}$$

$$\underline{g}_S = \sum_j \underline{F}^j \sum_I c_I \sigma(S,jI) + \sum_{jkl} \underline{I}^{jkl} \sum_I c_I \alpha_{lk}(S,jI) + \sum_T \underline{c}_T \gamma(S,T)$$

$$+ \underline{F} \sum_T \underline{c}_T \sigma(S,T) + \sum_{kl} [\underline{J}^{kl} \sum_T \underline{c}_T \alpha_{lk}(S,T) - \underline{K}^{kl} \sum_T \underline{c}_T \beta_{lk}(S,T)]$$

$$+ 2^{1/2} \sum_P p\ \{ \sum_j [\underline{C}_P \cdot \underline{F}^j + \underline{K}(\underline{C}_P)^j] \sigma(P,jS) + \underline{C}_P \cdot \sum_{jkl} \underline{I}^{jkl} \alpha_{lk}(P,jS) \} \ , \tag{34}$$

$$g_I = \sum_S \{ \sum_j \underline{c}_S^+ \cdot \underline{F}^j \sigma(S,jI) + \sum_{jkl} [\underline{c}_S^+ \cdot \underline{I}^{jkl}]\ \alpha_{lk}(S,jI) \}$$

$$+ 8^{-1/2} \sum_P \sum_{kl} tr[\underline{C}_P^+ (\underline{K}^{kl} + p\ \underline{K}^{lk})]\ \sigma(P,klI) \ . \tag{35}$$

In these equations $\underline{F}^j$ and $[\underline{K}(\underline{C}_P)]^j$ denote the j-th columns of the matrices $\underline{F}$ and $\underline{K}(\underline{C}_P)$, respectively, and $\underline{f} \cdot \underline{c}^+$ denotes the dyadic product of the vectors $\underline{f}$ and $\underline{c}$. The matrices $\underline{D}_P$ are defined as

$$(\underline{D}_P)_{ab} = (\underline{C}_P)_{ab} \ , \qquad (\underline{D}_P)_{ij} = 0 \ ,$$

$$(\underline{D}_P)_{ja} = p\ (\underline{D}_P)_{aj} = 2^{-1/2} \sum_S (\underline{c}_S)_a\ \sigma(P,jS) \ . \tag{36}$$

In this way all contributions of the integrals with three or four external orbitals to the matrices G_p are included in the operators $K(D_p)$. The operators $K(C_p)$ appearing in the expression for g_S can also be replaced by $K(D_p)$ if the coupling coefficients $\alpha_{kl}(S,T)$ and $\beta_{kl}(S,T)$ are modified (for details see Ref. 9). Hence, only one external exchange operator per pair per iteration has to be evaluated. The coupling coefficients σ, α, β, and γ are overlap integrals, first order transition density matrix elements or Hamiltonian matrix elements between the N-1 and N-2 electron core functions, e.g.,

$$\sigma(S,T) = \sum_{\sigma} \langle \Phi_{S\sigma} | \Phi_{T\sigma} \rangle \;, \tag{37}$$

$$\alpha_{kl}(S,T) = \sum_{\sigma} \langle \Phi_{S\sigma} | n_k^{\alpha^+} n_l^{\alpha} + n_k^{\beta^+} n_l^{\beta} | \Phi_{T\sigma} \rangle \;, \tag{38}$$

$$\beta_{kl}(S,T) = \langle \Phi_{S\alpha} | n_k^{\alpha^+} n_l^{\alpha} | \Phi_{T\alpha} \rangle + \langle \Phi_{S\beta} | n_k^{\beta^+} n_l^{\beta} | \Phi_{T\beta} \rangle$$
$$+ \langle \Phi_{S\alpha} | n_k^{\beta^+} n_l^{\alpha} | \Phi_{T\beta} \rangle + \langle \Phi_{S\beta} | n_k^{\alpha^+} n_l^{\beta} | \Phi_{T\alpha} \rangle \;, \tag{39}$$

$$\gamma(S,T) = \sum_{\sigma} \langle \Phi_{S\sigma} | H | \Phi_{T\sigma} \rangle \;. \tag{40}$$

Similiar expressions for the other coupling coefficients can be found in Ref. 9. The energy denominators in Eqs. 17 and 18 are given by:

$$\langle \Psi_S^a | H-E | \Psi_S^a \rangle = \left[\underline{F} + \sum_{kl} (\underline{J}^{kl} \alpha_{lk}(S,S) - \underline{K}^{kl} \beta_{lk}(S,S)) \right]_{aa} + \gamma(S,S) - E \tag{41}$$

$$E_P^{ab} = \varepsilon_P^a + \varepsilon_P^b + \left[(aa|bb) + p(ab|ab) \right]/(1+\delta_{ab}) - E \tag{42a}$$

$$\varepsilon_P^a = \left[F + \sum (J^{kl} \alpha_{lk}(P,P) - K^{kl} \beta_{lk}(P,P)) \right]_{aa} + \gamma(P,P)/2 \tag{42b}$$

Since the integrals $(aa|bb)$, $(ab|ab)$ are unavailable if a full integral transformation is not performed, we approximate

$$\left[(aa|bb) + p(ab|ab) \right]/(1+\delta_{ab}) \approx \frac{1}{4} \left[(\underline{J}^{ii} + \underline{J}^{jj})_{aa} + (\underline{J}^{ii} + \underline{J}^{jj})_{bb} \right] \tag{42c}$$

where the indices i,j label the core function Φ_{ijpm} that dominates Φ_{Pm}. This approximation might influence the speed of convergence but not the final solution.

It is important to note that expressions (33)-(35) are still valid if non-orthogonal configurations sets $\{\Psi_S^a\}$ and $\{\Psi_I\}$ are used. In fact, it is very advantageous to use the non-orthogonal configurations, since their internal structure is much simpler than that of the orthogonalized ones. This reduces the number of coupling coefficients and the computational effort for their evaluation. If the g_I and g_S are calculated in the non-orthogonal set they have to be transformed to the orthogonal basis in order to evaluate the matrix elements in Eqs. (31) and (32) and to update the coefficients c_I and c_S. The updated coefficients are then transformed back to the non-orthogonal basis and used to evaluate new g_I, g_S, and G_p. Since the overlap matrices given in Eqs. (11) and (12) are usually strongly blocked, these transformations require only negligible time. The N-2 electron functions Φ_{Pm} are orthogonalized, since otherwise more than one external exchange operator per pair per iteration would be required.

IV. TECHNICAL ASPECTS AND VECTORIZATION

a) Coupling coefficients. Due to the fact that we fully exploit the concept of the first order interacting space and that furthermore we internally contract the configurations, the core functions Φ_{Pm} and $\Phi_{S\sigma}$ are in general not eigenfunctions of the spin operator S^2. In our present program the core functions are therefore simply represented as linear combinations of determinants. Each determinant is stored as one integer word, whose bit pattern specifies the occupancy. The coupling coefficients can then be quite easily calculated using Boolean algebra. Our first brute force procedure was found to be satisfactorily efficient on a scalar machine (UNIVAC 1100/61) in which integer and logical operations are relatively fast compared to double precision floating point operations. After transfering the program to a CRAY-1 computer, we were faced with the problem that the calculation of the coupling coefficients, which is hardly vectorizable, became faster only by a factor of ≈ 12. Due to the efficient vectorization of the iteration procedure, however, the times for evaluating the eigenvector decreased by a factor of 200 or more (depending of the basis set size and the number of pairs). Hence, in our present implementation of the program on the CRAY, the calculation of the coupling coefficients sometimes dominates the entire calculation, particularly in cases with many reference determinants. Clearly, this is a very unsatisfactory situation.

In principle, the core functions could also be represented as linear combinations of particular spin-eigenfunctions, so that efficient, more advanced techniques like GUGA (21) could be applied for calculating the coupling coefficients. Probably one would have to evaluate them first in the special uncontracted configuration basis and then transform them to our contracted basis. Although this procedure seems to be feasable, it appears to be relatively complicated and it might also be quite I/O intensive.

For the present we have therefore decided to remain with the present procedure but to improve it as much as possible. In fact, there are several possibilities which should considerably reduce the computation times. First, each contracted core function may be considered as a linear combination of several "configurations", where a configuration here denotes a particular orbital occupancy irrespective of its spin coupling. These configurations are the same for all m or σ components (some components may vanish, however). By representing each configuration by one of its determinants in which all open-shell β-spin orbitals have been converted to α-spin orbitals, it is possible to precheck easily if a particular configuration pair gives a contribution. Test calculations have shown that in this way more than 80 per cent of all determinant matchings can be skipped. Furthermore it is already possible to determine on the configuration level which particular orbital pair k,l can give a non-vanishing matrix element in Eqs. (38), (39) etc., and also to determine the phase. Hence, most of the logic can be removed from the innermost loops in which the determinants are matched. Since one configuration is typically represented by 4-7 determinants, this should lead to a considerable saving. A second possibility to reduce the effort is to divide the valence orbital space into a closed-shell part and an open-shell part. The closed-shell part comprises the orbitals which are doubly occupied in all reference configurations. It is obvious that many of the coupling coefficients between core functions which have holes in the closed-shell part are identical. Calculating the coupling coefficients for each of these classes only once should save much time in cases with many closed-shell orbitals. Unfortunately, we are still in the process of implementing these techniques, and at present it is not yet possible to draw final conclusions about their efficiency.

b) Ordering of summations. The efficiency of a direct CI program may depend considerably on the order in which the summations in Eqs. (33)-(35) are performed. For the case of general uncontracted MR-CI wave functions, which are not restricted to the first order interacting configuration space, this has recently been discussed extensively by Saunders and van Lenthe (12). These authors conclued that it may even be useful to implement several alternative procedures into one program, and to determine automatically which one leads to the smallest number of floating point operations for a particular interaction. An important aspect in their considerations was the fact that there may be many spin couplings resulting from a given orbital occupancy. The number of configurations which differ only in the spin coupling of their internal orbitals is very much reduced, however, if only the first order interacting configuration space is taken into account. Furthermore, other algorithms in some cases appear to be optimal for internally contracted MR-CI wavefunctions rather than for uncontracted ones. For instance, in the P-P interactions for internally contracted wavefunctions, there are several operators $\underline{J}^{kl}$ and $\underline{K}^{kl}$ which couple a pair P with a different pair Q, whereas there is at most one operator of each type for uncontracted wavefunctions. Therefore, for internally contracted wave functions the minimal number of matrix multiplications is usually obtained if the operators $\underline{J}$ and $\underline{K}$ which contribute to a given interaction are contracted first (partial Hamiltonian matrix scheme in Ref. 12). For uncontracted wave functions, however, it may be more efficient to contract first all $\underline{C}_Q$ which contribute with a fixed operator $\underline{J}$ or $\underline{K}$ to a particular $\underline{G}_P$ (internal spin driven algorithm in Ref. 12). Similiar considerations apply to other types of interactions. In our program the summations are performed essentially in the order which is indicated in Eqs. (33)-(35) (rightmost summations or summations in brackets first).

c) Vectorization. In order to obtain a high efficiency on a CRAY-1 computer it is of utmost importance to convert as many vector operations as possible into matrix multiplications. This is due to the fact that for matrix multiplications the maximal Mflop rate (millions of floating point operations per second) is about three times larger (≈ 150) than for vector additions $\underline{a}=\underline{a}+fak\cdot\underline{c}$ (≈ 50 Mflop). For instance, the times for the S-S interactions could be reduced by about a factor of three by expanding the non-zero elements $\alpha_{kl}(S,T)$ (all T for given S,kl) into a (eventually sparse) vector $a(T)$ and then performing the vector times matrix multiplication $v(a) = v(a) + \Sigma_T a(T)\cdot c(T,a)$. Using an assembler (CAL) routine written by Saunders the vanishing elements in $a(T)$ can be skipped without losing much efficiency. In a similiar way the integral vectors $\underline{I}^{jkl}$ appearing in the P-S interactions are contracted using an expanded coefficient vector $a(jkl) = \alpha_{kl}(P,jS)$ (fixed P,S). In order to perform the matrix multiplications most efficiently, symmetric or antisymmetric matrices stored in triangular form have to be expanded to square form. The time for these operations is small, and it does not seem necessary to keep the full matrices on mass storage devices. A crucial part of any direct CI calculation is the evaluation of the external exchange operators $\underline{K}(\underline{D}_P)$. On a CRAY, these operators can certainly most efficiently be calculated as proposed by Saunders and van Lenthe (12), i.e., by storing the integrals $(ac/bd)\pm(ad/bc)$ as 64×64

blocked square super-matrices (ab=columns, cd=rows). However, this requires storing most integrals twice on the disk. We have chosen a somewhat different algorithm which is probably somewhat slower but avoids the duplication of the integrals. The integrals are ordered into triples (ab/cd), (ac/bd), and (ad/bc) for a>b>c>d. After reading a buffer of integrals, we copy them into three triangular matrices $(\underline{J}^{ab})_{cd}$, $(\underline{K}^{ab})_{cd}$, and $(\underline{K}^{ba})_{cd}$. Appropriate linear combinations $\underline{J}\pm\underline{K}$ are then formed and expanded to square form. Finally, the contributions are obtained simultaneously to as many operators as possible in terms of matrix times vector multiplications, where the vector length is determined by the number of pairs per symmetry. This algorithm does not require to expand the matrices $\underline{C}_p$ to square form.

d) Use of sparseness of $(\Delta\underline{C}_p)$. The computational effort for calculating the matrices $\underline{G}_p$ can be considerably reduced by evaluating $\Delta\underline{G}_p$ instead of $\underline{G}_p$ in each iteration. Since all elements in $(\Delta\underline{C}_p)_{ab}$ which are smaller than about $5 \cdot 10^{-6}$ can usually be neglected without significant loss of accuracy, the matrices $\Delta\underline{C}_p$ become very sparse at the end of the iteration process (see Fig.1a). Using the outer product algorithm the effort for the matrix multiplications in Eq. (33) can then be reduced considerably (see Fig. 1b). Provided the two-electron integrals are fully

Fig.1a 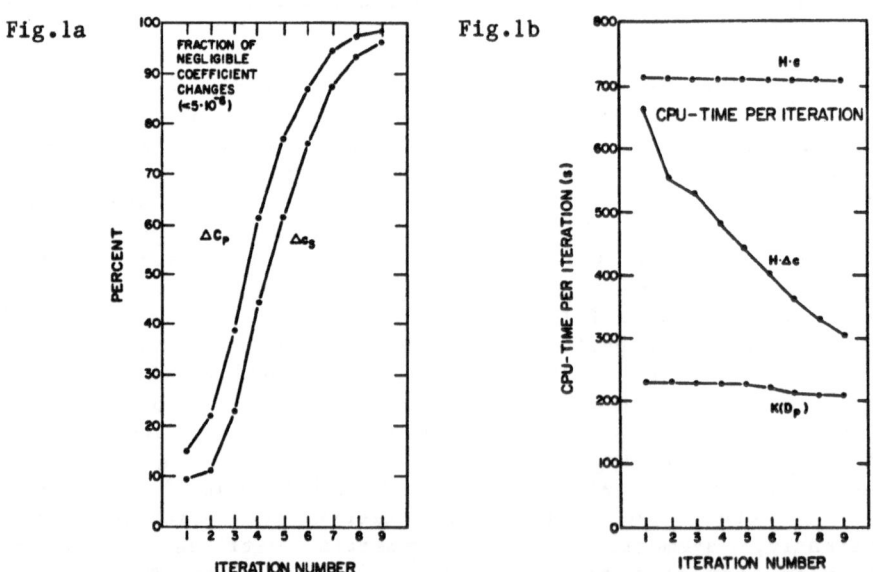 Fig.1b

Fig. 1a: Fraction of negligible coefficient changes $\Delta\underline{c}_S$ and $\Delta\underline{C}_p$ for H_2O; 69 orbitals, 11 reference configurations.

Fig. 1b: Decrease of computation time per iteration with increasing sparsity of coefficient vectors and matrices for the example in Fig. 1a. Timings are for UNIVAC 1100/61 double precision.

transformed into the MO basis, the timings for calculating the $K(D_P)$ can be reduced correspondingly, but this is not the case in our present program. On a scalar machine it is also useful to employ the Δc_S instead of c_S, even though IF statements are required in some of the innermost loops. For instance, on a UNIVAC 1100/61 we have found that the times for the operations a=a+fak·b are reduced if b contains more than about 20 per cent zeros. On a vector machine, however, it does not appear worthwhile to check for zero elements in Δc_S.

V. TEST CALCULATIONS AND APPLICATIONS

In this section we first present some typical timings for SCEP calculations on a CRAY-1 computer. Secondly, some internally contracted and uncontracted MCSCF-SCEP calculations will be compared. Finally, a brief overview of recent applications of the method will be presented.

In Table 1 some details of test calculations on ozone are listed. These calculations were performed in order to compare with timings recently published by Saunders and van Lenthe (SVL) (12). The basis set and SCF energies were exactly the same in both cases. In the CI calculations of SVL, three energetically high lying

Table 1

Comparison of SD-CI calculations for ozone[a]

	$^1A_1$		$^3B_1$	
	SCEP	SVL[b]	SCEP	SVL[b]
Number of configurations	35279	31440	41482	123393
CPU-times (sec CRAY-1):				
integral transformation	3.4[c]	49[d]	3.6[c]	49[d]
time per iteration	3.3	2.4	4.7	11.1
number of iterations	11	11	7	13
total time	39.7	75	32.8	144
correlation energy	-0.58195	-0.58191	-0.50417	-0.50494

a) Basis [641], 69 contractions, see Ref.12.
b) Saunders and van Lenthe, Ref.12.
c) Partial integral transformation, not fully vectorized.
d) Full integral transformation, no symmetry used.

orbitals have been omitted. Therefore, the number of configurations for the closed-shell singlet state is somewhat smaller in their calculation than in our SCEP calculation, but this has only a negligible effect on the correlation energy. For the triplet state, however, the number of configurations in the SVL calculation is about three times larger than that in the SCEP wave function. This is due to the fact that SVL consider all possible singly and doubly excited configurations, whereas in the SCEP calculation only the first order interacting configuration space has been taken into account. As seen from Table 1, the omission of the non-interacting configurations degrades the correlation energy only by 0.15 per cent, but makes the calculation much cheaper. It is interesting to note that the number of iterations needed to converge the triplet state was considerably smaller for the SCEP calculation then for the SVL calculation, even though the Davidson procedure has been employed only in the latter case (same accuracy in both cases). Presumably, this is due to the fact that the non-interacting configurations converge slowly. By comparing the timings in Table 1, it must be noted that SVL have first performed a full integral transformation. This transformation was relatively expensive, since it has been done without using symmetry. The transformation times could certainly be reduced by a factor of four at least by fully

Table 2

SCEP-CEPA calculations^a on $[PtCl_4]^{2-}$

State	$^1A_{1G}$	$^3B_{1G}$
number of valence orbitals	20	21
number of pairs/singles/internals	400/20/1	438/819/55
number of configurations (C_{2v})	153266	172908
time for $\underline{J}^{kl}, \underline{K}^{kl}$	7.0	7.8
time per iteration	22.2	38.3
relative timings (in percent):		
P-P interactions	91.5	59.7
S-S interactions	0.6	7.9
P-S interactions	4.1	22.6
S-I interactions	0.6	7.6
other	3.2	2.2
operators $\underline{K}(\underline{D}_p)$	29.3	18.5

a) Triple zeta valence basis (75 contractions); core electrons treated by effective potentials. Times are CPU-seconds on CRAY-1.

Table 3

Details of internally contracted MCSCF-SCEP calculations for CN $A^2\Pi$
(74 orbitals, 19 reference configurations, 60 pairs, 293 singles,
162 internals, 266841 configurations, 37704 variational parameters)

Relative CPU-times in per cent (CRAY-1):

	c_I	$\underline{c}_S$	$\underline{c}_P$
g_I	1.8	5.0	0.8
$\underline{g}_S$	4.3	12.1	11.3
$\underline{G}_P$	1.8	11.4	51.7

Total CPU-time per
iteration was 7.5 sec.

Relative time for $\underline{K}(\underline{D}_P)$
was 23.6 per cent.

exploiting the symmetry. As seen by comparing the data for the
singlet state, the time per iteration was somewhat smaller in SVL's
calculation than in ours. A comparison more detailed than that
done in Table 1 shows that this is solely due to the difference in
treating the contributions of integrals with three and four
external orbitals. In SVL's method, these contributions were
evaluated from the transformed integrals, whereas in our present
SCEP program the operators $\underline{K}(\underline{D}_P)$ were calculated directly from
the integrals in the AO basis. This is more expensive, since a
larger number of integrals must be processed. The transformations
of the matrices $\underline{D}_P$ into the AO basis and the back transformations
of the $\underline{K}(\underline{D}_P)$ into the MO basis take less than 10 per cent of the
total time needed for the operators and is not the main reason for
the difference in the computation times. From the data in Table 1
it is obvious that for a calculation of this size it is not
worthwhile to perform the full integral transformation. Even if the
symmetry were fully used, the savings in CPU time would be small if
any. This is in contradiction to the conclusions of SVL.

Generally, it is to be expected that the cost effectiveness of
a full integral transformation increases with increasing number of
pairs. Table 2 shows some data for larger calculations with up to
438 pairs. Even in this case the times for evaluating the operators
$\underline{K}(\underline{D}_P)$ are relatively short, and to perform a full integral
transformation would certainly save very little time. It is
interesting to note that the times for the S-S interations in the
calculation of the triplet state are short, although there are 819
vectors $\underline{c}_S$. On the other hand, the P-S interactions are

Table 4

Number of configurations and variational parameters in various
internally contracted MCSCF-SCEP calculations[a]

molecule	number of ref.conf.	number of orbitals valence	total	number of config.	var.param.
O_3	2	10	69	69764	41378
OH	7	7	52	33803	13267
CH_2 $(^3B_1)+H_2$	5	6	68	55425	19703
H_2O	11	7	69	74372	20893
SiO	8	8	78	135948	40650
$CN(A^2\Pi)$	19	8	74	266841	37676
N_2 $(B^2\Pi)$	21	8	90	637524	64347

a) Only C_{2v} symmetry used in all cases; the CPU time per iteration
on a CRAY-1 for the contracted N_2 calculations was 22.0 sec. The
CPU times per iteration for the uncontracted and contracted H_2O
calculations on a UNIVAC 1100/61 were 41.4 and 15.0 min,
respectively. The corresponding timings for CH_2+H_2 were 43.5 and
14.0 min on a DEC-10 computer (double precision arithmetic).

relatively expensive, which is due to the fact that the evaluation
of the dyadic products required in this part has not yet been fully
vectorized. Using assembler subroutines for the innermost loops the
times   for   this   part   and   also   for   the   partial   integral
transformation (operators $\underline{J}$, $\underline{K}$) could certainly be reduced by a
factor of three or more.

Table 3 shows timings for an internally contracted MCSCF-SCEP
calculation with a fairly complex reference function. The relative
times for the different interactions are found to be quite similiar
to those for the $^3B_{1G}$ state of $[PtCl_4]^{2-}$. The total time per
iteration is only slightly larger than the time needed for the
triplet state of O_3 (cf. Table   1) with a single reference
determinant and about the same number of variational parameters.

In Table 4 the numbers of configurations and variational
parameters   for   various   internally   contracted   MCSCF-SCEP
calculations are listed. For the cases shown, the internal
contraction reduces the number of parameters typically by a factor
between 3 and 10. The computation times per iteration are reduced
by about the same factors (see examples in footnote of Table 4).
The savings become larger with increasing complexity of the
reference wave functions. The largest reduction of the number of
parameters is found in cases where the reference wavefunction
contains no closed-shell valence orbitals. This is the case for the
N_2 calculations, where the number of parameters is reduced by a
factor of 10.

Table 5

Comparison of correlation energies[a] for internally contracted and uncontracted MCSCF-SCEP wave functions

molecule	number of ref.conf.	correlation energy contracted	correlation energy uncontratced	difference (per cent)
$OH(X^2\Pi)$[b]	3	-0.2168	-0.2171	0.14
	7	-0.2214	-0.2218	0.18
$CH_2(^3B_1)+H_2$[c]	5	-0.1688	-0.1691	0.18
$O_3(^1A_1)$[b,d]	2	-0.6192	-0.6208	0.26
$H_2O(^1A_1)$[b]	11	-0.2684	-0.2691	0.23

a) $E_{korr}=E_{tot}-E_{SCF}$; b) near equilibrium geometry; c) at saddle point geometry of the reaction $CH_2+H_2 \rightarrow CH_3+H$, see Ref.7 ; d) the coefficients of the reference configurations are:

MCSCF	0.88192	-0.47140
contracted CI	0.91117	-0.41668
uncontracted CI	0.92616	-0.38863

In Table 5 a comparison of correlation energies computed with internally contracted and uncontracted MCSCF-SCEP wave functions is presented. In most cases the contraction degrades the correlation energies by less than 0.25 per cent. The largest difference is found for O_3. This is expected, however, since in this case there are two strongly mixing states, whose relative weights in the wave-function considerably change as the electrons are correlated (cf. coefficients below Table 5). Even then the error due to the contraction amounts to only 0.26 per cent and is completely negligible compared to other errors, such as those introduced by incompleteness of the basis set or the neglect of further reference configurations.

Table 6 demonstrates that the internal contraction also has a negligible effect on other properties such as dipole or electronic transition moments. The contraction changes the moments by less than 0.5 per cent in all cases. The largest absolute error amounts to 0.01 Debye and is observed for the A state at R=2.5 Bohr. This error is still smaller than the uncertainties due to other approximations in the wave function, which for calculations of this quality typically amount to 0.02 Debye for a diatomic hydride. It should be noted that for the calculations in Table 6 the same set of molecular orbitals has been employed for both states, since this considerably facilitates the calculation of the electronic transition moments. The orbitals were obtained by minimizing the energy average of both reference functions (23,24). Other test

Table 6

Comparison of dipole and transition moments[a] of $OH(X^2\Pi - A^2\Sigma^+)$ for internally contracted and uncontracted wave functions[b]

| | R=1.8 Bohr | | | R=2.5 Bohr | | |
	μ_1	μ_2	μ_{12}	μ_1	μ_2	μ_{12}
contracted	0.6505	0.6540	0.1405	0.6598	0.9941	0.02991
uncontracted	0.6495	0.6518	0.1412	0.6585	0.9900	0.02981
difference (%)	0.15	0.33	0.50	0.28	0.41	0.33

a) in a.u.; b) 8 and 10 reference configurations for X and A states respectively. The orbitals used for both states have been obtained by minimizing the energy average of the two reference functions.

calculations on OH have shown that the errors of the dipole moments due to the internal contraction are even smaller if the reference wave functions are individually optimized (9).

The accuracy of computed potential energy functions is most conveniently checked by comparing the computed spectroscopic constants with experimental data. In Table 7 such a comparison is made for a number of diatomic molecules which were recently investigated in our laboratory (25-31). In all cases relatively large, flexible Gaussian basis sets have been employed. The equilibrium distances are mostly too long by 0.003-0.005 Å, which is most likely due mainly to basis set defects. For the case of C_2^- only, the calculated distances are in error by 0.008 Å, which can be attributed to the fact that only 2 d-functions and no f-functions have been included in the basis set used for the carbon atom. The harmonic constants ω_e are generally in excellent agreement with experiment. The typical error for the energy difference $\Delta G_{1/2}$ of the two lowest vibrational levels amounts to 10-15 cm^{-1} (<1%). For the case of OH the potential energy function has also been calculated with uncontracted MCSCF-SCEP wavefunctions. The deviations of the spectroscopic constants obtained from the contracted and the uncontracted calculations are negligible (9).

From the results in Table 7 it is obvious that accurate potential energy functions can often be obtained using relatively simple reference wavefunctions which just describe dissociation properly. Unfortunately, this does not appear to be the case for one-electron properties such as dipole moments. Figure 2 shows the results of an investigation of the stability of dipole moments of OH with respect to improvements of the reference wave function. In order to obtain stable results, it turned out to be important to

Table 7

Comparison of calculated and experimental spectroscopic constants[a]

molecule	Ref.	number of ref.conf.	r_e	B_e	α_e	ω_e	$\omega_e x_e$
BH	34	3	1.235	12.00	0.413	2362	47
			1.232	12.02	0.412	2366.9	49.4
OH	9,25,27	3 / 3[b]	0.971	18.84	0.728	3737	88
			0.971	18.83	0.726	3736	87
			0.970	18.91	0.724	3735	84.9
SH	31	3	1.340	9.61	0.269	2734	50
			1.341	9.60	(0.270)	(2711)	(59)
OH^+	25	7	1.031	16.71	0.744	3108	82
			1.029	16.79	0.749	3113.3	78.5
SiO	26	8	1.515	0.722	0.005	1242	5.8
			1.510	0.727	0.005	1241.5	5.97
C_2^- $(X^2\Sigma^+)$	28	10	1.276	1.726	0.016	1780	12
			1.268	1.747	0.0166	1781.3	11.7
$(A^2\Pi)$		14	1.318	1.619	0.016	1646	11
			(1.33)	1.58±0.1		(1646)	(10.5)
$(B^2\Sigma^+)$		12	1.231	1.855	0.017	1983	16
			1.223	1.877	0.0173	1969.1	14.9
CN $(X^2\Sigma^+)$	29	12	1.174	1.892	0.017	2079	13
			1.172	1.900	0.0174	2068.7	13.1
$(A^2\Pi)$		19	1.238	1.700	0.017	1796	12
			1.233	1.716	0.0172	1813	12.3

a) internally contracted MCSCF-SCEP calculations for molecular ground states if not otherwise stated; experimental values given in last line for each state. Uncertain experimental values in brackets; r_e in Å, all other values in cm^{-1}; so far, potential energy functions of similiar accuracy have also been obtained for $OH^-(X^1\Sigma^+)$ [25], $OH(A^2\Sigma^+)$, $HF^+(X^2\Pi, A^2\Sigma^+)$, $HCl^+(X^2\Pi, A^2\Sigma^+)$ [27], $N_2(A^3\Sigma_u^-, B^3\Pi_g, W^3\Delta_u, B'^3\Sigma_u^-)$ [30]. b) uncontracted MCSCF-SCEP.

include in the reference function configurations accounting for $\pi \rightarrow \pi^*$ excitations. It should be noted that the dipole moments in Fig. 2 have been calculated by numerical differentation of the total energy with respect to an applied external electric field. If the dipole moments are calculated as expectation values over the CI wavefunctions, they are somewhat more sensitive with respect to variations of the reference wavefunction, in particular at the larger internuclear distance. For details of these investigations the reader is refered to the original paper (25).

In a recent thorough CASSCF (complete active space MCSCF) Larsson and Siegbahn (32) found that the transition moment for the $CH(A^2\Delta-X^2\Pi)$ transition is very slowly convergent with respect to the number of active orbitals and to the size of the basis set. Furthermore, early multireference CI calculations of Hinze, Lie and Liu (3) yielded lifetimes which are in poor agreement with recent accurate experimental values (33,34). We therefore found it worthwhile to perform a few test calculations to check the accuracy achieved with MCSCF-SCEP wavefunctions. The results of this investigation are presented in Table 8 and compared graphically with the CASSCF data in Fig. 3. On the abscissa of

Fig. 2   Fig. 3

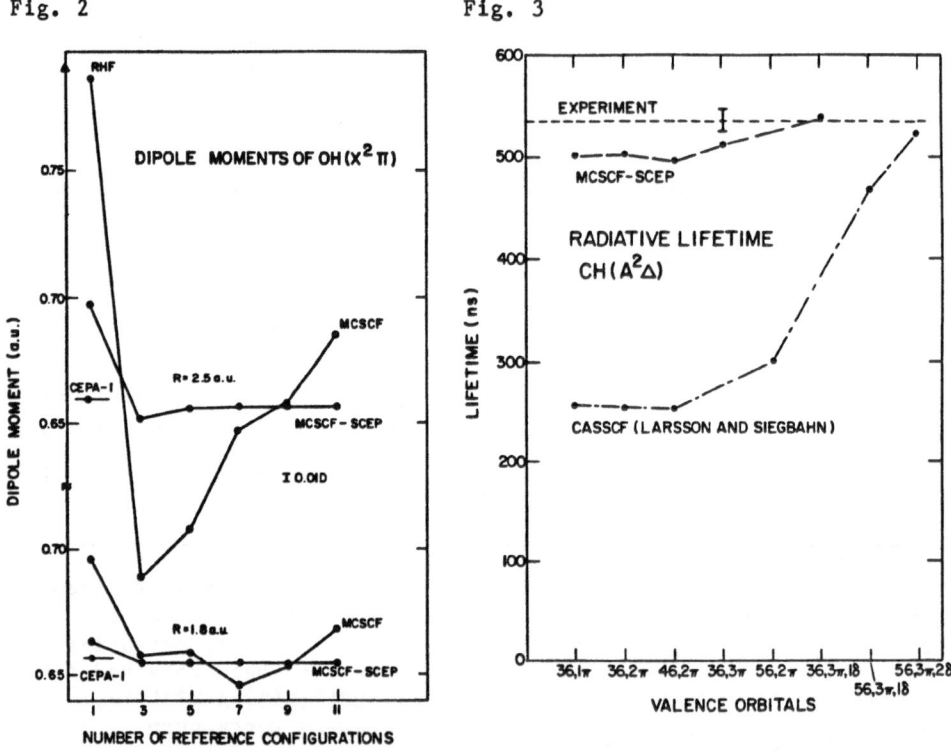

Fig. 2: Dependence of calculated dipole moments of OH on the number of reference configurations (for details see Ref.25).

Fig. 3: Dependence of radiative lifetimes of CH $A^2\Delta$ on the number of active orbitals in CASSCF calculations (from Ref. 31) and on the number of valence orbitals in the MCSCF-SCEP calculations. The basis set used in the CASSCF calculations is not the same in all cases, but this does not change qualitatively the results.

Table 8

MCSCF-SCEP calculations of transition moments for CH $(X^2\Pi - A^2\Delta)$

ref.conf.		valence	basis	transition-	lifetime	energy
$X^2\Pi$	$A^2\Delta$	orbitals	set	moment(a.u.)	(ns)	diff.(eV)
4	5	31[b]	A[c]	0.282	502[d]	3.01
8	7	32	A	0.280	505	3.00
9	10	33	A	0.278	512	3.00
11	10	42	A	0.283	495	3.00
10	11	331	A	0.270	543	3.00
10	11	331	B	0.272	538	2.93
10	11	331	C	0.273	531	2.91
Experiment					536±10	2.87

a) at R=2.1 Bohr; b) 331 means 3σ, 3π, and $1\delta_{xy}$ valence orbitals (1σ electrons not correlated); c) basis A: 12s,8p,4d - 7s,3p; basis B: 12s,8p,4d,1f-7s,3p,1d; basis C: 12s,8p,4d,2f-7s,3p,1d. d) Lifetimes have been estimated by scaling the lifetimes of Larsson and Siegbahn(31) by $[\mu_{12}(ref)/\mu_{12}(calc)]^2$.

Figure 3 those orbitals which comprise the active orbital space in the CASSCF calculations and the valence orbital space in the SCEP calculations are specified. It is found that even with the smallest reference wave function our calculated lifetime deviates by less than 7 per cent from the experimental values. These CI calculations are certainly much cheaper than the large scale CASSCF calculations required to achieve the same level of accuracy. In order to approach the experimental value more closely, it turned out to be most important to include a δ_{xy} orbital into the reference wave function. The same has been found by Larsson and Siegbahn in their CASSCF calculations. This seems, however, to be an exceptional case, since in similar investigations of CN (29) and N_2 (30) the inclusion of δ-orbitals into the reference function had only a very small effect on the transition moments. It is interesting to note that the excitation energies in Table 8 are very insensitive to the variations of the reference wave functions. The remaining errors seem to be mainly due to basis set deficiencies, as indicated by the improvements of the excitation energies by adding f-functions to the basis set.

Finally, in Table 9 we present a summary of radiative lifetimes of electronically excited states which have recently been calcula- ted from internally contracted MCSCF-SCEP wave functions (27-30). For a detailed discussion of these results we refer to the original

Table 9

Comparison of calculated[a] and experimental radiative lifetimes of electronically excited states (in μs unless otherwise stated)

Molecule/State	MCSCF-SCEP	Experiment	References(33-53)
OH $A^2\Sigma^+$, v'=0	590 ns	625±25 ns	German et al.,1973
		688±21 ns	German, 1975
		788±13 ns	Brophy et al.,1976
		760±20 ns	Brozowski et al.,1978
		686±14 ns	Dimpfl and Kinsey, 1979
		720±9 ns	Dermid, Laudenslager,1982
HF$^+$ $A^2\Sigma^+$,v'=0	26.2	7.2±0.7	van Sprang, de Heer, 1978
HCl$^+$ $A^2\Sigma^+$,v'=0	2.51	2.6±0.2	Moehlmann et al., 1977
		3.4±0.4	Martner et al.,1983
CH $A^2\Delta$,v'=0	531 ns	534±5 ns	Brozowski et al.,1976
		537±5 ns	Becker et al., 1980
C_2^- $B^2\Sigma_u^+$,v'=0	76.5 ns	77±8 ns	Leutwyler et al.,1982
CN $A^2\Pi$,v'=0	11.0	14.2	Sneden et al.,1982
		7.3±1.5	Conley et al.,1981
v'=10	5.3	5.3	Sneden et al.,1982
N_2 $B^3\Pi_g$,v'=0	12.0	8.0±1	Jeunehomme, 1966
v'=5	6.5	6.2±0.6	Jeunehomme, 1966
		2.4	Hartfuss et al.,1968
		6.1±0.2	Hollstein et al.,1969
		3.5	Johnson et al.,1970
		5.1±1	Chen and Anderson,1975
		5.4±0.2	Eyler and Pipkin, 1983
$W^3\Delta_u$,v'=1	1.13 ms	2±1 ms	Covey et al.,1973
v'=5	104	93±50	Covey et al.,1973

a) internally contracted MCSCF-SCEP calculations. The accuracy of the theoretical lifetimes is estimated to be 10-15 per cent.

papers. As seen in Table 9, the experimental data obtained by different authors are often in quite poor agreement; generally the deviations of different measurements are much larger than the error bounds given for the individual values. Therefore, it appears impossible to draw general conclusions about the accuracy of our calculated values by comparing them with the experimental data. However, various test calculations similiar to those in Table 8 have been performed in order to check the stability of our results with respect to variations of the reference configurations and the basis sets. In these tests the transition moments were always stable within about two per cent. Although it cannot be completely ruled out that other variations have a larger effect on the results, we feel that it can be quite safely assumed that the calculated lifetimes are accurate within 10-15 per cent. Thus, it appears that, at present, radiative lifetimes of small molecules can be more reliably calculated than measured. In any case the calculations are much cheaper than measurements of the same accuracy.

Acknowledgements

The authors like to thank Prof. W. Meyer for numerous valuable discussions on the subject. One of us (HJW) also likes to thank P.J. Hay for his hospitality and cooperation at the Los Alamos National Laboratory. HJW also gratefully acknowledges a grant from the Deutsche Forschungsgemeinschaft.

References

1. For a review see, e.g., I. Shavitt, in "Modern Theoretical Chemistry", edited by H.F.Schaefer III, Plenum, New York 1977
2. R.J. Buenker, S.D. Peyerimhoff and W. Butscher, Mol. Phys. 35, 771 (1978); for a recent review see R.J. Buenker and S.D. Peyerimhoff, in "New Horizons of Quantum Chemistry", edited by P.-O. Loewdin and B. Pullmann, D. Reidel, Dordrecht, 1983
3. G.C. Lie, J. Hinze, and B. Liu, J.Chem. Phys. 59, 1872,1887 (1973); J. Hinze, G.C. Lie, and B. Liu, Astrophys, J. 196, 621 (1975) S. Chu, M. Yoshimine, and B. Liu, J. Chem. Phys. 61, 5389 (1974)
4. B.R. Brooks and H.F. Schaefer III, J. Chem. Phys. 70, 5092 (1979); B.R. Brooks, W.D. Ladig, P.Saxe, N.C. Handy, and H.F. Schaefer III, Phys. Scr. 21, 312 (1980)
5. W. Meyer, Int. J. Quant. Chem. 5, 341 (1971); J. Chem. Phys. 58, 1017 (1973); Theor. Chim.Acta 35, 277 (1974)
6. P.R. Taylor, J. Chem. Phys. 74, 1256 (1981)
7. P.E.M. Siegbahn, J. Chem. Phys.70, 5391 (1979); 72, 1647 (1980)
8. H. Lischka, R.Shepard, F.B. Brown, and I.Shavitt, Int.J.Quant. Chem. S15, 91 (1981)
9. H.-J. Werner and E.-A. Reinsch, Proceeedings of the "5 th Seminar on Computational Methods in Quantum Chemistry", Groningen 1981, edited by T.H. van Duijen and W.C. Nieuwpoort, Max Planck Inst., Garching, Muenchen; J. Chem. Phys. 76, 3144 (1982).
10. B. Liu and M. Yoshimine, J. Chem. Phys. 74, 612 (1981).
11. P.Saxe, D. J. Fox, H. F. Schaefer III, and N. C. Handy, J. Chem. Phys. 77, 5584 (1982).
12. V. R. Saudners and J.H.van Lenthe, Mol. Phys. 48, 923 (1983).
13. I. Shavitt, contribution to this volume.

14. A. Bunge, J. Chem. Phys. 5, 20 (1970)
15. A.D.McLean and B.Liu,J. Chem. Phys. 58, 1066 (1973)
16. W. Meyer, in "Modern Theoretical Chemistry", see Ref. 1
17. P.E.M. Siegbahn, Chem. Phys. 25, 197 (1977); "Proceedings of the 5th Seminar on Computational Methods in Quantum Chemistry", see Ref. 9.
18. W. Meyer, J. Chem. Phys. 64, 2901 (1976)
19. R. Ahlrichs, in "Proceedings of the 5th Seminar on Computational Methods in Quantum Chemistry", see Ref. 9.
20. E.R. Davidson, J. Comp. Phys. 17, 87 (1975)
21. I. Shavitt, in "New Horizons of Quantum Chemistry" (see Ref. 2) and references therein.
22. H.-J. Werner, J. Chem. Phys., submitted
23. K.K. Docken and J. Hinze, J. Chem. Phys. 57, 4928 (1972)
24. H.-J. Werner and W. Meyer, J. Chem. Phys. 74, 5784 (1981)
25. H.-J. Werner, P. Rosmus and E.A. Reinsch, J. Chem. Phys. 79, 905 (1983)
26. H.-J. Werner, P. Rosmus and M. Grimm, Chem.Phys. 73, 169 (1982)
27. H.-J. Werner, P. Rosmus, W. Schaetzl, and W. Meyer, J. Chem. Phys., in press
28. P. Rosmus and H.-J. Werner, J. Chem. Phys., submitted
29. H.-J. Werner, P.J. Hay, and D. Cartwright, J. Chem. Phys., submitted.
30. P. Kalcher, H.-J. Werner and E.-A. Reinsch, in preparation
31. H.-J. Werner, unpublished results
32. M. Larsson and P.E.M. Siegbahn, J. Chem. Phys. 79, 2270 (1983)
33. J. Brzozowski, P. Bunker, N.Elander, and P. Erman, Astrophys. J. 207, 414 (1976)
34. K.H.Becker, H.H. Brenig, and T. Tatarczyk, Chem. Phys. Lett. 71, 242 (1980)
35. K.R. German, T.H. Bergeman, E.M. Weinstock, and R.N. Zare, J. Chem. Phys. 58, 4304 (1973)
36. K.R. German, J. Chem. Phys. 62, 2584 (1975); 63, 5252 (1975)
37. J.H. Brophy, J.A. Silver, J.L. Kinsey, Chem. Phys. Letters 28, 418 (1976)
38. J.Brzozowski, P. Erman, and M.Lyrra, Phys. Scr. 17, 507 (1978)
39. W.L. Dimpfl and J.L. Kinsey, J. Quant. Spectr. Rad. Trans. 21, 233 (1979)
40. I.S. McDermid and J.B. Laudenslager, J. Chem. Phys. 76, 1824 (1982)
41. H.A. van Sprang and F.I. de Heer, Chem. Phys. 21, 73 (1978)
42. G.R. Moehlmann, K.K. Bhutani, and F.I. de Heer, Chem. Phys. 21, 127 (1978)
43. C. Martner, J. Pfaff, H. Rosenbaum, A.O'Keefe, and J. Saykally, J. Chem. Phys. 78, 7074 (1983)
44. S. Leutwyler, J.P. Meier, and L. Misev, Chem. Phys. Lett. 91, 206 (1982)
45. C. Sneden and D.C. Lambert, private communication
46. C. Conley, J.B. Halpern, J. Wood, C. Vaughn, and W.M. Jackson, Chem. Phys. Lett. 73, 274 (1980)

47. M. Jeunehomme, J. Chem. Phys. 45, 1805 (1966)
48. H.J. Hartfuss and A. Schmillen, Z. Naturforsch. Teil A23, 722 (1968) 49. M. Hollstein, D.C. Lorentz, J.R. Peterson, and J.R. Sheridan, Can. J. Chem. 47, 1858 (1969)
50. A.W. Johnson and R.G. Fowler, J. Chem. Phys. 53, 65 (1970)
51. S.T. Chen and R.J. Anderson, Phys. Rev. A12, 468 (1975)
52. E.E. Eyler and F.M. Pipkinm, J. Chem. Phys, 79, 3654 (1983)
53. R. Covey, K.A. Saum, and W. Benesch, J. Opt. Soc. Am. 63, 592 (1973)

COMPUTATIONAL ASPECTS OF DIRECT SCF AND MCSCF METHODS

Jan Almlöf* and Peter R. Taylor†

*Department of Chemistry, University of Oslo,
 Blindern, Oslo 3, Norway.
†CSIRO Division of Chemical Physics, P.O. Box 160,
 Clayton, Victoria, 3168, Australia.

INTRODUCTION

In quantum chemical research a considerable fraction of the
effort is devoted to improving the computational strategy. The
work in this field can basically be characterized as aiming at a
single goal: how can larger systems be made more accessible to
increasingly sophisticated methods? It is evident that the
development of modern computer hardware and software has greatly
accelerated progress towards this goal. This development, however,
has not been uniform: rather, different functions of the average
computer have improved in performance by different degrees. In
order to design optimally efficient computational methods, such
changes in computer architecture and performance profile must be
carefully analyzed.

Virtually all large-scale quantum chemical calculations make
the following use of computational facilities:
1) Very CPU-intensive activity, often array-oriented, requiring
   a minimum of 8-10 decimal digits of accuracy;
2) Storage - temporary or permanent - of large amounts of data.
   Normally this data is used in each cycle of an iterative
   process, and data retrieval will place a heavy load on the
   I/O system of the installation.
In many types of quantum chemical calculation, the data handling
referred to in 2) above is not an intrinsic part of the
mathematical method. On the contrary, the whole philosophy of
saving and retrieving large amounts of data from one iteration to
another was initially justified on the grounds that CPU activity
was slow compared to I/O, and that arithmetic floating-point
operations constituted the real bottleneck. With the hardware

C. E. Dykstra (ed.),
Advanced Theories and Computational Approaches to the Electronic Structure of Molecules, 107–125.
© *1984 by D. Reidel Publishing Company.*

available today this is not necessarily the case. In fact, the
practical limitation on many types of calculation is not so much
the CPU time requirements for computing and processing the data,
but rather the external storage needed and/or the I/O time spent
reading the data in each iterative cycle. From the trends seen
currently in the development of new computer hardware, there are
reasons for believing that this situation will be accentuated in
the future.

   With the hardware and software available today, systems with
more than 200-300 basis functions seem to represent the practical
upper limit, even in conventional single configuration Hartree-
Fock work using large mainframe systems. Obviously, alternative
approaches to large-scale *ab initio* LCAO calculations are called
for. The solution to be discussed here is quite unsophisticated
from a conceptual viewpoint - it amounts simply to avoiding some
or all of the I/O and storage requirements by recalculating data
whenever it is needed, rather than retrieving it from external
memory. Such a scheme was originally suggested and designed in
order to make large SCF calculations possible on minicomputer
systems: the procedure was termed "Direct SCF" [1], and it has
made it possible to study systems with 400-500 basis functions [2].
Clearly, the evaluation of two-electron integrals over a basis of
such a size presents a computational challenge by itself,
particularly if the calculation is to be carried out on a mini.
This holds *a fortiori* if the integrals are to be recalculated in
every iteration, and a few requirements are rather obvious:

A)   The algorithms and code used for the evaluation of integrals
      must be the fastest available. This problem has received
      considerable attention in recent years [3-6], and quite
      efficient schemes are now available.
B)   The number of times the integrals are recomputed, i.e. the
      number of iterations, must be kept as small as possible. As
      the total integral time will be proportional to the number
      of iterations, it is evidently justifiable to spend more
      effort on improving convergence than is the case for
      conventional calculations. Schemes that provide exact or
      approximate quadratic convergence have been suggested [7,8]
      and should have great advantages when used with direct SCF
      approach.
C)   In any iteration, the amount of data to be recomputed should
      be kept to a minimum, with the additional constraint that
      this does not increase I/O demand on the system.

   The requirements A and B above apply also to conventional
LCAO-SCF calculations, and it is not surprising that substantial
effort has been spent in these areas with such applications in
mind. Their importance and relative weight in the direct SCF
scheme is, however, quite different.

In the case of post-Hartree-Fock methods, both MCSCF and CI, the appropriate equation systems are conventionally written in terms of integrals over MOs and in computational implementations an integral transformation is performed. For large basis sets the I/O overheads then become even greater than for SCF, and it is certainly of interest to try to devise "direct" schemes to alleviate the I/O problem. Various possible schemes are discussed in detail elsewhere [9], with particular reference to CI calculations; we consider below several aspects pertinent to the MCSCF case.

In the following sections we shall discuss "direct evaluation" as opposed to "retrieval from external memory", both for the SCF and MCSCF case. Particular emphasis will be placed on possibilities for exploiting the potential of vector-oriented hardware in such calculations.

DIRECT SCF CALCULATIONS - GENERAL CONSIDERATIONS

A conventional LCAO Hartree-Fock calculation can be logically and technically divided into two different steps. In the first step the necessary basis set integrals over the Hamiltonian are computed and stored. In the second, the Roothaan-Hall equations are solved iteratively. Computationally, the most demanding part of this step amounts to reading the full list of integrals generated in the first step and processing these to form the Fock matrices required in the Roothaan procedure. In a closed-shell calculation, the contributions from a two-electron integral $[\mu\nu|\lambda\sigma]$ are given by

1) Coulomb contributions:

$$F_{\mu\nu}: = F_{\mu\nu} + 4D_{\lambda\sigma}[\mu\nu|\lambda\sigma] \tag{1a}$$

$$F_{\lambda\sigma}: = F_{\lambda\sigma} + 4D_{\mu\nu}[\mu\nu|\lambda\sigma] \tag{1b}$$

2) Exchange contributions:

$$F_{\mu\lambda}: = F_{\mu\lambda} - D_{\nu\sigma}[\mu\nu|\lambda\sigma] \tag{1c}$$

$$F_{\nu\sigma}: = F_{\nu\sigma} - D_{\mu\lambda}[\mu\nu|\lambda\sigma] \tag{1d}$$

$$F_{\mu\sigma}: = F_{\mu\sigma} - D_{\nu\lambda}[\mu\nu|\lambda\sigma] \tag{1e}$$

$$F_{\nu\lambda}: = F_{\nu\lambda} - D_{\mu\sigma}[\mu\nu|\lambda\sigma] \tag{1f}$$

In the early days of *ab initio* calculations the computational effort expended in calculating the integrals in the first step dominated the entire calculation. However, with the developments in efficient algorithms which have occurred, this is no longer the case when Gaussian basis sets are used. The time required for

building the Fock matrix or matrices may be 5-20% of the time
required to compute the integrals, and the Fock matrix build must
be repeated in each iteration; it is thus clear that factors
other than the integral time alone must be considered in selecting
an optimum strategy. Such considerations led some time ago to the
concepts of $\mathcal{P}$- and $\mathcal{K}$-supermatrices:

$$\mathcal{P}(\mu\nu|\lambda\sigma) = 4[\mu\nu|\lambda\sigma] - [\mu\lambda|\nu\sigma] - [\mu\sigma|\nu\lambda] \tag{2a}$$

$$\mathcal{K}(\mu\nu|\lambda\sigma) = [\mu\lambda|\nu\sigma] + [\mu\sigma|\nu\lambda] \tag{2b}$$

With these constructs the Fock matrix build is considerably
simplified. Using folded triangular density matrices the closed-
shell case illustrated above reduces to

$$F_{\mu\nu}: = F_{\mu\nu} + D_{\lambda\sigma}\mathcal{P}(\mu\nu|\lambda\sigma) \tag{3a}$$

$$F_{\lambda\sigma}: = F_{\lambda\sigma} + D_{\mu\nu}\mathcal{P}(\mu\nu|\lambda\sigma) \tag{3b}$$

The time used to perform the transformation (2) is small when
efficiently coded, and this overhead is easily recouped when (3)
is used in each iteration instead of (1).

Similar reductions may be achieved by exploiting the point
group symmetry of the molecule. It is evident that symmetry will
cause relations among the integrals in addition to those arising
from trivial index permutational symmetry. Using these relations
will reduce the time spent in the calculation, as fewer unique
integrals need to be calculated, thus immediately reducing the
integral evaluation time. If the Fock matrix is constructed with
contributions from every symmetry-related integral in (1) or (3)
the SCF time will be roughly the same as without symmetry, and
therefore will dominate even more than in a case without symmetry.
As shown by Dacre [10] and Elder [11], and further developed by
Dupuis and King [12], the rate-determining Fock matrix build can
be performed using only symmetry-unique integrals - the time ratio
integrals/SCF will then be roughly the same irrespective of whether
symmetry is used or not.

A further step towards reduction of the integral list may be
taken by transforming the integrals to a symmetry-adapted basis.
Such a transformation is generally a time-consuming affair.
However, in the special case of point groups with real one-
dimensional irreducible representations (i.e. D_{2h} and its subgroups) the transformation may be carried out with particular
efficiency, representing point group operators and irreducible
representations by bit strings and translating the group
theoretical manipulations into elementary Boolean algebra. The
time used in such a transformation will be proportional to N^4m^2,
whereas the number of unique integrals (and hence the integral

time) goes as N^4m^{-1}, N and m being the number of basis functions
and the order of the group, respectively. A symmetry transformation does not automatically reduce the number of integrals to be
stored or processed in a calculation. In the case of an SCF
calculation, however, only integrals involving symmetry orbitals
from at most two irreducible representations are needed, and when
supermatrices are used the number of elements to be saved behaves
as N^4m^{-2}, while the symmetry transformation time behaves as N^4m.
The comparison between the different approaches to symmetry is
complicated by the fact that many integrals are small enough to
be neglected in a normal calculation: these terms are omitted
from the integral list and do not contribute to either the storage
requirements or the processing time. Unfortunately, both the
supermatrix technique and the use of symmetry-adapted integrals
tend to delocalize the electronic interaction and consequently
the resulting lists will have fewer negligible elements.

CONTRACTION

The segmented contraction scheme (i.e. any primitive function
contributing to only one contracted function) seems to present a
problem on modern vector hardware, in the sense that efficient
vectorization is inhibited. The general contraction scheme [13]
is much more favourable in this respect, however, care must still
be taken in order to achieve the optimum performance. Apparently,
a four-index n^5 transformation scheme would involve the minimum
number of operations. On the other hand, the index range in each
step of the procedure will be too short to permit efficient
vectorization. The transformation should therefore be arranged
so that in any quarter-transformation the loops over the indices
involved in the transformation are performed outside the loops
over the three passive indices. A straightforward way to achieve
this is to perform the transformation steps

$$[\alpha\beta|\gamma\delta] \to [\mu\beta|\gamma\delta] \to [\mu\nu|\gamma\delta] \to [\mu\nu|\lambda\delta] \to [\mu\nu|\lambda\sigma] \qquad (4)$$

using as index quadruples

$$(\alpha,\beta,\gamma,\delta) , (\beta,\gamma,\delta,\mu) , (\gamma,\delta,\mu,\nu) ,$$

$$(\delta,\mu,\nu,\lambda) , (\mu,\nu,\lambda,\sigma) \qquad (5)$$

respectively, where $\alpha,\beta,\gamma,\delta$ represent primitive functions, μ,ν,λ,σ
contracted functions, and the index $(\alpha,\beta,\gamma,\delta)$ denotes storage
with the first index running fastest, etc. If n_1, n_2, etc.
primitives are involved and m_1, m_2, etc. contracted functions, we
may define a compressed passive index from the first three indices
of a quadruple in (5), say for $(\beta,\gamma,\delta,\mu)$ as

$$p = \gamma + (\delta-1)*n_3 + (\mu-1)*n_3*n_4 \qquad (6)$$

Given the running order of the various indices mentioned above, the location for the integrals $[\mu\beta|\gamma\delta]$ and $[\mu\nu|\gamma\delta]$ involved in this particular quarter-transformation is given by

$$\beta + (p-1)*n_2 \tag{7a}$$

and

$$p + (\nu-1)*n_3*n_4*m_1 \tag{7b}$$

Clearly, these integrals may be accessed in a regular way, with the innermost loops over μ, γ and δ collapsed into a single loop over p. Organized in this way a general contraction scheme would vectorize efficiently on a computer where contiguous vectors are not necessary, such as the CRAY. The procedure can then be implemented with vector lengths ranging between $m_1*m_2*m_3$ and $n_2*n_3*n_4$.

It should be noted that the number of floating-point operations required in a general contraction scheme is quite large. As an example, it would usually take fewer operations to set up the Fock matrix in the primitive basis and then to transform it to the contracted basis. This involves a one-electron transformation only, and such a procedure will always be preferable when more than two contracted functions are constructed from a given set of primitives. For conventional LCAO calculations, in which contraction is performed in order to reduce storage requirements, rather than to save on arithmetic operations, such arguments are irrelevant. In the direct SCF approach, however, storage considerations apply only to the extent that the one-electron matrices in an uncontracted basis may become too large to fit in the available memory - if the required matrices can be accommodated it is preferable in the direct case to avoid contraction entirely. An additional advantage is that a more accurate and less subjective result is thereby obtained; considerable overhead can also be eliminated from the code.

FOCK MATRIX CONSTRUCTION

The evaluation of primitive integrals can be carried out rather efficiently on a vector processor, as discussed in depth by a number of authors [3,5], and we shall not consider this aspect further here. The subsequent step in a direct SCF iteration is the Fock matrix construction, which merits further consideration. The supermatrix formalism (3) is not attractive for direct SCF, since the transformation (2) would have to be carried out in each iteration. Accordingly, we focus our interest on how Fock matrices may be built efficiently using (1) in the case of recomputation of integrals in each iteration. Clearly, if parallelism in the integral evaluation is exploited a a batch of integrals will be available simultaneously for processing. The batch may involve different components of four shells (intrinsic vectorization) or functions of the same angular type but with different combinations of orbital exponents (extrinsic vectorization). The problem of vectorizing (1) obviously lies in ordering Fock and density matrices such that their elements may be accessed in a regular order for all the operations (1a-1f) as the program loops over integrals in a batch.

One way to achieve the necessary regularity would be to perform two GATHER operations and one SCATTER on the matrices for each step (1a) to (1f). In most cases, however, it would be more efficient to order the basis functions such that the required matrix elements appear at regular intervals in the matrices. This requirement will be fulfilled if all functions of the same quantum number and type are grouped together. Using the extrinsic schemes, the vectorization should be performed on one of the four indices looping over all basis functions of a given type. With the grouping of basis functions described above this would yield acceptable vector lengths on most machines. If shell structure is to be exploited in the integral program, which is usually desirable, it may be necessary to process the integrals further and to increment the Fock matrix inside the shell loop in order to reduce memory requirements. The choice between these alternatives may be made during execution with negligible overhead.

INTEGRAL PRE-SCREENING

In any computational work, it is customary, before time-consuming sub-tasks are performed, to test whether the results of the sub-task will significantly affect the final result. The extent to which such pre-testing is worthwhile depends on the efficiency with which the test can be performed, the frequency of a positive outcome of such a test, the complexity of the operations to be avoided and on possible degradation of program efficiency as a result of implementing such testing. In conventional LCAO calculations, such testing is performed in the evaluation of two-electron integrals. Integrals are computed in batches associated with a quadruplet of shells. Within a batch, the product of radial overlaps

$$T_{\mu\nu\lambda\sigma} = S_{\mu\nu}S_{\lambda\sigma} \tag{8a}$$

is a common factor which largely determines the magnitude of all integrals in the batch. By comparing this factor against a threshold,

$$T_{\mu\nu\lambda\sigma} < \varepsilon \tag{8b}$$

a single test may eliminate the evaluation of an entire block.

On a vector processor one might want to avoid this individual
testing in order not to inhibit vectorization. Similar advantages
may be obtained, however, by using appropriate COMPRESS/EXPAND or
GATHER/SCATTER techniques based on a vector of the radial overlap
products $T_{\mu\nu\lambda\sigma}$.

While these general considerations on testing apply to both
direct SCF and to conventional integral evaluation procedures,
there are additional simplifications to be exploited in the direct
SCF case. Since the usage of an integral is fully known at any
stage of the integral evaluation, further possibilities for
testing whether a given integral will make a significant
contribution exist. From (1) it is evident that the maximum
contribution $[\mu\nu|\lambda\sigma]$ makes to any Fock matrix element is given by

$$[\mu\nu|\lambda\sigma].\max(4|D_{\lambda\sigma}|,4|D_{\mu\nu}|,|D_{\mu\lambda}|,|D_{\nu\sigma}|,|D_{\mu\sigma}|,|D_{\nu\lambda}|) \qquad (9a)$$

Since the density matrix is known prior to integral evaluation,
the test variable can be modified accordingly, suggesting the use
of

$$T'_{\mu\nu\lambda\sigma} = T_{\mu\nu\lambda\sigma}.\max(4|D_{\lambda\sigma}|,4|D_{\mu\nu}|,|D_{\mu\lambda}|,|D_{\nu\sigma}|,|D_{\mu\sigma}|,|D_{\nu\lambda}|) \quad (9b)$$

which would exclude more integrals than (8b). This idea can be
carried a step further [14]: noting that the contribution to the
total energy from the above integral is given by

$$[\mu\nu|\lambda\sigma].(4D_{\mu\nu}D_{\lambda\sigma} - D_{\mu\lambda}D_{\nu\sigma} - D_{\mu\sigma}D_{\nu\lambda}) \qquad (10a)$$

the test criterion can be modified to

$$T''_{\mu\nu\lambda\sigma} = T_{\mu\nu\lambda\sigma}.\max(4|D_{\mu\nu}D_{\lambda\sigma}|,|D_{\mu\lambda}D_{\nu\sigma}|,|D_{\mu\sigma}D_{\nu\lambda}|) \qquad (10b)$$

This test would eliminate more integrals than (9b), as the
product of density matrix elements is small when either of the
elements is small. Unfortunately, our numerical experience with
respect to the convergence behaviour of such calculations is
discouraging. In closed-shell systems, however, a similar approach,
which is more efficient than (9) but which retains the convergence
properties of using the exact Fock matrix, may be used. We note
first that the convergence problems associated with (10b) occur
because the procedure does not guarantee any particular accuracy
in the Fock matrix elements. If defects in accuracy occur in
Fock matrix elements coupling MOs of different type (i.e. one MO
occupied, the other empty), the density matrix to be used in the
next iteration will be affected to first order, and convergence
problems are then inevitable. Conversely, inaccuracies may be
tolerated in Fock matrix elements which couple MOs of the same
type. In order to implement this we note first that the contribution from a two-electron integral in the AO basis to a Fock matrix

element in the MO basis is given by

$$F_{ia} := F_{ia} + [\mu\nu|\lambda\sigma] \cdot (R_{\mu\nu\lambda\sigma} - \tfrac{1}{4}R_{\mu\lambda\nu\sigma} - \tfrac{1}{4}R_{\mu\sigma\nu\lambda}) \tag{11a}$$

where

$$R_{\mu\nu\lambda\sigma} = Q_{\mu\nu}D_{\lambda\sigma} + Q_{\lambda\sigma}D_{\mu\nu} \tag{11b}$$

and

$$Q_{\mu\nu} = C_{\mu i}C_{\nu a} + C_{\nu i}C_{\mu a} \tag{11c}$$

Here, $C_{\mu i}$ and $C_{\nu a}$ are coefficients for the ith occupied and ath unoccupied MO, respectively. This suggests the following modified test criterion for the integral evaluation:

$$T_{\mu\nu\lambda\sigma} \cdot \max(4B_{\mu\nu\lambda\sigma}, B_{\mu\lambda\nu\sigma}, B_{\mu\sigma\nu\lambda}) \tag{12a}$$

where

$$B_{\mu\nu\lambda\sigma} = (E_{\mu}G_{\nu} + E_{\nu}G_{\mu}) \cdot |D_{\lambda\sigma}| + (E_{\lambda}G_{\sigma} + E_{\sigma}G_{\lambda}) \cdot |D_{\mu\nu}| \tag{12b}$$

and

$$E_{\mu} = \max|C_{\mu i}| \quad , \quad G_{\mu} = \max|C_{\mu a}| \tag{12c}$$

The integral pre-screening can also be made more efficient by utilizing the information retained in the Fock matrix from the previous iteration. A change in the Fock matrix from iteration (m-1) to iteration (m) can be written and computed as

$$\delta \underline{\underline{F}}(m) = \oint \delta \underline{\underline{D}}(m-1) \tag{13}$$

where $\delta\underline{\underline{D}}(m-1)$ is the increment in the density matrix when going from iteration (m-2) to (m-1). All of the previous suggested test criteria may be based on $\delta\underline{\underline{D}}$ rather than on $\underline{\underline{D}}$. This normally leads to a more stringent test, as only changes in the density call for a re-evaluation of integrals. As convergence is reached the number of integrals to be calculated will approach zero.

Processing the integrals in the batches determined by shell structure introduces some additional considerations in the use of these test criteria. Clearly, in a shell-oriented approach it is of little value to know if individual integrals in the batch will give a negligible contribution, as the efficiency of the technique depends on treating all integrals in the batch equivalently. In order for the test to be "safe" with respect to overall accuracy, the entire batch should be evaluated whenever any integral in the batch gives a significant contribution. A contracted density matrix may therefore be constructed as

$$D_{MN} = \max \left| D_{\mu\nu} \right| \tag{14}$$

where μ and ν are individual basis functions within the shells M and N. This contracted density matrix is constructed at the beginning of each iteration, and is then used in expressions (9), (10) or (12) for tests pertaining to the entire batch of integrals.

TECHNICAL LIMITATIONS

In the direct SCF approach, the bottleneck of external storage has been removed. As a consequence, the size of the systems which can be studied increases drastically, and other limitations begin to require attention. Thus a calculation with 700 AOs would need some 4 Mbytes for storing one Fock/density matrix pair during Fock matrix construction. Hence memory requirements during this phase may be the factor ultimately limiting the size of system which can be studied with LCAO methods. Of course, this particular limitation is not unique to the direct SCF approach, but it is usually obscured by other more serious restrictions in the conventional scheme.

Virtual memory management techniques in their current form offer no immediate solution to the problem. On a reasonably fast computer about 10 000 integrals can be computed every second (all overhead included), and every integral is to be multiplied by six density matrix elements and added to six Fock matrix elements, all of which are more or less randomly distributed in memory. At present, it is beyond the capacity of any virtual memory system to comply with page requests at such a rate. Fast SSD memories may provide a solution to the problem, albeit at considerable cost currently. Computational methods which reduce the memory requirements represent an alternative avenue which certainly deserves investigation, and we discuss here one scheme which may be used for systems with symmetry.

Conventional symmetry blocking of the Fock and density matrices is not practical, as a symmetry transformation of the integrals would be required. However, when symmetry is present there will be redundancies among integrals and among density matrix elements, as outlined above in the discussion of symmetry, consequently, it is unnecessary to store all density and Fock matrix elements. In the Dacre-Elder scheme only a "skeleton" Fock matrix is computed during integral processing; this skeleton matrix is subsequently "symmetrized" to give the full matrix [12]. It is straightforward to organize processing so that only this skeleton matrix need be stored, as follows.

A loop over all basis functions can be decomposed into the following structure:

```
  - Loop over symmetry independent shells M
  -- Loop over operators R generating symmetry-related shells RM
  --- Loop over components in the shell RM
```

The reduction of the density matrix to a symmetry redundant set is then simply achieved by omitting the loop over symmetry operations R in the generation of the first AO index. That is, batches of density matrix elements D(M,RN) are computed, where M and N index symmetry-independent shells and R runs over all operators of the point group. Similar restrictions can be employed in the calculation of two-electron integrals [15], requiring only the batches [M RN|SΛ $T\Sigma$]. Processing these integrals requires density matrix elements from D(M,RN), D(M,$S\Lambda$), D(M,$T\Sigma$), D(RN,$S\Lambda$), D(RN,$T\Sigma$) and D($S\Lambda$,$T\Sigma$). Of these, only the first three are available from the symmetry-reduced list held in memory. The remaining three, however, can be obtained from the list from the equivalent transformed values D(N,$R^{-1}S\Lambda$), D(N,$R^{-1}T\Sigma$) and D(Λ,$S^{-1}T\Sigma$). Similarly, the Fock matrix contributions from these terms can be transformed back to batches in which the first index refers to the symmetry independent set, so that the contributions are added directly to the symmetry-reduced skeleton Fock matrix. While this strategy does not produce the same skeleton Fock matrix as does the original Dacre-Elder procedure, the correct Fock matrix is obtained after symmetrization.

DIRECT MCSCF - GENERAL CONSIDERATIONS

We shall restrict discussion here to MCSCF wave functions which are used to overcome inadequacies of the SCF model, such as near-degeneracy effects. MCSCF optimization of SDCI expansions and the like is discussed elsewhere [9]. The calculation involves an MO space with is partitioned into _inactive_ MOs i,j..., doubly occupied in all MCSCF configuration state functions (CSFs), _active_ MOs t,u..., showing variable occupancy among the MCSCF CSFs, and _secondary_ MOs a,b..., unoccupied in all CSFs. The type of MCSCF wave function to be considered is characterized by a small active MO space, of dimension 2 - 20, say. The MCSCF wave function |0> can be expanded in the CSFs |K> as

$$|0> = \sum_K |K>c_K \qquad (15)$$

Given some guess at the MOs ϕ and expansion coefficients $\underline{c}$ the full second-order Newton-Raphson equation system for the MCSCF optimization can be written as

$$\begin{bmatrix} \underline{\underline{E}}^{xx} & \underline{\underline{E}}^{xc} \\ \underline{\underline{E}}^{cx} & \underline{\underline{E}}^{cc} \end{bmatrix} \begin{bmatrix} \underline{X} \\ \underline{\delta c} \end{bmatrix} = \begin{bmatrix} \underline{E}^{x} \\ \underline{E}^{c} \end{bmatrix} \qquad (16)$$

Here the updated parameters are given by

$$\underline{c}' = \underline{c} + \underline{\delta c} \tag{17}$$

$$\underline{\phi}' = \exp(\underline{\underline{X}})\underline{\phi} \tag{18}$$

and the arrays of derivatives in (16) are given by

$$(\underline{\underline{E}}^{X})_{pq} = \frac{\partial E}{\partial X_{pq}} \tag{19}$$

$$(\underline{\underline{E}}^{XC})_{pq,K} = \frac{\partial^2 E}{\partial X_{pq} \partial c_K} \tag{20}$$

evaluated at $\underline{X} = \underline{0}$ and at the estimated coefficients $\underline{c}$. Note that the vector $\underline{X}$ is simply the lower triangle of the antisymmetric matrix $\underline{\underline{X}}$. Expressions for the derivatives, in terms of MO integrals, the effective one-electron operator

$$F_{pq} = h_{pq} + \sum_i \{2[ii|pq] - [ip|iq]\} \tag{21}$$

and reduced density matrices over the wave function (15) have been given by a number of authors (see e.g. refs 16-18 and refs therein). In summary, $\underline{E}^c$, $\underline{E}^{cc}$, $\underline{E}^{xc}$, $\underline{E}^{cx}$ and $\underline{E}^x$ involve only $[tu|vx]$, $[tu|vi]$ and $[tu|va]$ and elements of $\underline{F}$ from (21); $\underline{E}^{xx}$ also involves $[ij|ab]$, $[ia|jb]$, $[tu|ab]$, $[ta|ub]$, $[ij|tu]$, $[it|ju]$, $[ia|jt]$ and $[ia|tb]$. This suggests that, in addition to the Fock matrix-like construction of $\underline{F}$, we are required to obtain $\underline{\underline{J}}^{ij}$, $\underline{\underline{K}}^{ij}$, $\underline{\underline{J}}^{it}$, $\underline{\underline{K}}^{it}$, $\underline{\underline{J}}^{tu}$ and $\underline{\underline{K}}^{tu}$, where

$$J^{ij}_{pq} = (p|J^{ij}|q) = [ij|pq] \tag{22}$$

and

$$K^{ij}_{pq} = (p|K^{ij}|q) = [ip|jq], \text{ etc.} \tag{23}$$

These requirements can, however, be considerably simplified when examined in the context of the method to be used to solve the equation system (16). A convenient scheme here is the iterative approach described by Pople and co-workers [19] for the solution of coupled perturbed Hartree-Fock equations; this involves expansion of the unknown $\underline{X}$ and $\underline{\delta c}$ in a set of basis vectors, and the main computational overhead is associated with multiplication of the Hessian matrix in (15) by these vectors. That is, it is necessary to form the product

$$\begin{bmatrix} \underline{\underline{E}}^{xx} & \underline{\underline{E}}^{xc} \\ \underline{\underline{E}}^{cx} & \underline{\underline{E}}^{cc} \end{bmatrix} \begin{bmatrix} \underline{a} \\ \underline{b} \end{bmatrix},$$

where $\underline{a}$ and $\underline{b}$ are expansion vectors for $\underline{X}$ and $\underline{\delta c}$, respectively. We can now eliminate the need to determine many of the $\underline{\underline{J}}$ and $\underline{\underline{K}}$ arrays by expanding terms in (24) explicitly.

Integrals such as $[ij|ab]$ and $[ia|jb]$ enter (24) only as a contribution to terms such as $\sum_{jb} E^{XX}_{ia,jb}\, a_{jb}$, in the form

$$\sum_{jb}(4[ia|jb] - [ib|ja] - [ij|ab])a_{jb} \tag{25}$$

$$= \sum_{jb\mu\nu\lambda\sigma}(4[\mu\nu|\lambda\sigma] - [\mu\lambda|\nu\sigma] - [\mu\sigma|\nu\lambda])a_{jb}C_{\mu i}C_{\nu a}C_{\lambda j}C_{\sigma b}$$

Defining

$$d_{\lambda\sigma} = \sum_{jb}a_{jb}C_{\lambda j}C_{\sigma b} \tag{26}$$

(25) becomes

$$\sum_{\mu\nu}P_{\mu\nu}C_{\mu i}C_{\nu a} \tag{27}$$

where

$$P_{\mu\nu} = \sum_{\lambda\sigma}(4[\mu\nu|\lambda\sigma] - [\mu\lambda|\nu\sigma] - [\mu\sigma|\nu\lambda])d_{\lambda\sigma} \tag{28}$$

Thus the contribution of the MO integrals $[ij|ab]$ and $[ia|jb]$ can be expressed completely in terms of AO integrals, as was shown for the SCF case by Bacskay [8].

Integrals such as $[ia|tb]$ appear in Hessian matrix elements such as $E^{XX}_{ia,vb}$ as

$$\sum_{t}D^{o}_{tv}(4[ia|tb] - [it|ab] - [ib|ta]), \tag{29}$$

where $\underline{D}^{o}$ is the first-order reduced density matrix given using the vector $\underline{c}$. Such Hessian matrix elements contribute to two types of $\underline{E}^{XX}\underline{a}$ product in (24). The first is $\sum_{vb}E^{XX}_{ia,vb}a_{vb}$, in which the integrals appear in the term

$$\sum_{tvb}D^{o}_{tv}(4[ia|tb] - [it|ab] - [ib|ta])a_{vb} \tag{30}$$

$$= \sum_{tvb\mu\nu\lambda\sigma}(4[\mu\nu|\lambda\sigma] - [\mu\lambda|\nu\sigma] - [\mu\sigma|\nu\lambda])a_{vb}D^{o}_{tv}C_{\mu i}C_{\nu a}C_{\lambda t}C_{\sigma b}$$

Defining

$$e_{\lambda\sigma} = \sum_{t}\sum_{b}a_{\sigma b}D^{o}_{tv}C_{\lambda t}C_{\sigma b}\,, \tag{31}$$

and

$$Q_{\mu\nu} = \sum_{\lambda\sigma}\sum (4[\mu\nu|\lambda\sigma] - [\mu\lambda|\nu\sigma] - [\mu\sigma|\nu\lambda])e_{\lambda\sigma}, \tag{32}$$

we obtain

$$\sum_{\mu}\sum Q_{\mu\nu}C_{\mu i}C_{\nu a} \tag{33}$$

for (30). The other type of product we must consider is $\sum_{ia}\sum E^{xx}_{ia,vb}a_{ia}$. The contribution is

$$\sum_{iat}\sum\sum D^0_{tv}(4[ia|tb] - [it|ab] - [ib|ta])a_{ia} \tag{34}$$

which can be rewritten as

$$\sum_{t\lambda\sigma}\sum\sum D^0_{tv}P_{\lambda\sigma}C_{\lambda v}C_{\sigma b} \tag{35}$$

using (26) and (28) above.

Finally, integrals such as $[it|ja]$ also contribute to two types of product in (24). The first is $\sum_{iv}\sum E^{xx}_{iv,ja}a_{iv}$, with contribution

$$\sum_{ivt}\sum\sum (2\delta_{tv}-D^0_{tv})(4[it|ja] - [ij|ta] - [ia|jt])a_{iv} \tag{36}$$

$$= \sum_{ivt\mu\nu\lambda\sigma}\sum\sum\sum\sum\sum\sum (4[\mu\nu|\lambda\sigma] - [\mu\lambda|\nu\sigma]$$

$$- [\mu\sigma|\nu\lambda])(2\delta_{tv}-D^0_{tv})a_{ib}C_{\mu i}C_{\nu t}C_{\lambda j}C_{\sigma a}.$$

Defining

$$f_{\mu\nu} = \sum_{ivt}\sum\sum (2\delta_{tv}-D^0_{tv})a_{iv}C_{\mu i}C_{\nu t} \tag{37}$$

and

$$R_{\lambda\sigma} = \sum_{\mu\nu}\sum (4[\mu\nu|\lambda\sigma] - [\mu\lambda|\nu\sigma] - [\mu\sigma|\nu\lambda])f_{\mu\nu}, \tag{38}$$

we obtain

$$\sum_{\lambda\sigma}\sum R_{\lambda\sigma}C_{\lambda j}C_{\sigma a} .$$

The other type of product is $\sum_{ja}\sum E^{xx}_{iv,ja}a_{ja}$, which can be simplified using $\underline{d}$ from (26) as was done in (34) - (35).

It is thus clear that with the aid of the "densities" $\underline{d}$, $\underline{e}$ and $\underline{f}$ and either the $\mathcal{P}$-supermatrix or the AO integral list the need to compute $\underline{J}^{ij}$, $\underline{K}^{ij}$, $\underline{J}^{it}$ and $\underline{K}^{it}$ can be eliminated completely. Obviously, once a formulation of this type has been arrived at, a "direct" implementation for many of the terms in (24) can follow the same lines as for direct SCF, although for (24) a number of density matrices must be handled simultaneously. Symmetry processing and pre-screening techniques may be applied as for SCF, and technical aspects of processing these terms are discussed below. However, no mathematical restructuring such as that performed above can eliminate the need for the arrays $\underline{J}^{tu}$ and $\underline{K}^{tu}$ (at least, not without creating substantial new computational difficulties), and we now discuss direct methods for generating these operator matrices.

DIRECT CONSTRUCTION OF OPERATOR MATRICES

Two possible schemes suggest themselves for a conventional approach to operator generation. The first is based on the simple updating procedure

$$[\mu\nu|tu]: = [\mu\nu|tu] + [\mu\nu|\lambda\sigma]c_{\lambda t}c_{\sigma u}, \tag{40}$$

followed by a one-electron transformation to give $\underline{J}^{tu}$. An analogous procedure is followed for $\underline{K}^{tu}$. The updating (40) behaves as $n_A^2 N^4$ for n_A active MOs. Alternatively, a standard N^5-type algorithm could be used, (40) being replaced by

$$[\mu\nu|\lambda u] = \sum_{\sigma}[\mu\nu|\lambda\sigma]c_{\sigma u} \tag{41a}$$

$$[\mu\nu|tu] = \sum_{\lambda}[\mu\nu|\lambda u]c_{\lambda t} \tag{41b}$$

The latter approach behaves as $n_A N^4$.

For a direct approach the implementation of (40) is very simple, being essentially equivalent to setting up multiple Fock matrices in direct SCF. This is also true for the $\underline{K}^{tu}$. (40) can handle AO integrals in arbitrary order, so that no sorting of the integrals is required. However, unless all $\underline{J}^{tu}$ and $\underline{K}^{tu}$ can be held in storage simultaneously, the AO integrals will have to be recomputed a number of times, this number increasing both as n^2 and as N^2 for fixed memory length.

The approach (41) is rather more difficult to implement in a direct scheme. It is necessary to set up $[\mu\nu|\lambda\sigma]$ for all $\lambda\sigma$ and as many $\mu\nu$ values as will fit in available storage: the transformations (41) are then performed and the half-transformed

integrals are written out for sorting. Setting up the $[\mu\nu|\lambda\sigma]$
in this way requires modification of the loop structure of a
conventional integral program; this becomes even more
complicated for the $\underline{K}^{tu}$ case. These new loop structures result
in the explicit computation of integrals whose contribution
would conventionally be obtained by the use of index permutation
symmetry. A further disadvantage is the need to perform a disk-
based sort. On the other hand, at least in a naive view, the AO
integrals need be generated only once.

An overall comparison of the advantages and disadvantages of
the two schemes is given elsewhere [9], and it is not necessary
to repeat the details here. Rather, we shall discuss some
computational considerations concerning both schemes, with
particular reference to vector processors, as was done for the SCF
case above.

DIRECT MCSCF - COMPUTATIONAL IMPLEMENTATION AND VECTOR PROCESSORS

The construction of the matrices $\underline{F}$, $\underline{P}$, $\underline{Q}$ and $\underline{R}$ in (21), (28),
(32) and (38) can be implemented exactly in the fashion outlined
for the Fock matrix in the direct SCF case, as outlined above.
All of the special features described earlier, such as integral
pre-screening, contraction after matrix construction and the
exploitation of symmetry, can be implemented in the same way.
The symmetry processing is exactly the same, as the expansion
vectors $\underline{a}$, and the transformation $\exp(\underline{X})$, only mix MOs which
transform according to the same row of the same irreducible
representation. Similarly, any discussion of Fock matrix
construction on vector processors applies directly in the present
case.

The $n_A^2 N^4$ and $n_A N^4$ transformation schemes behave quite
differently in the context of vectorization. The $n_A N^4$ approach
can readily be vectorized in terms of matrix multiplications of
length N, a task which will be performed fairly efficiently on
most vector processors, especially where N is fairly large
(>200, say). The $n_A^2 N^4$ scheme can easily be vectorized with
vector length n_A^2 (assuming n_A^2 J or K arrays can be held in memory
together) but this will generally be unsatisfactory for $n_A < 8$.
Again, it would be possible to vectorize processing on batches of
integrals, as in the SCF case above, but this might lead to
extensive data motion or the need to handle a complicated indexing
scheme. The question of pre-screening in the two approaches is
considered below.

The evaluation of batches of integrals, in order to maximize
efficiency in the integral evaluation, poses special problems for
the $n_A N^4$ scheme. In intrinsic vectorization of the integrals, for
J operators, say, it will be necessary to hold arrays $[\mu\nu|\lambda\sigma]$ for

all $\lambda\sigma$ and for all $\mu\nu$ arising from a particular pair of shells M, N, if integral recomputation is to be avoided. This may not be possible for M, N high angular momentum shells and large AO basis sets. Analogously, if extrinsic vectorization is used for the integral computation it will be necessary to hold $[\mu\nu|\lambda\sigma]$ for all $\lambda\sigma$ and for all $\mu\nu$ arising from functions of the same angular type but with different exponents. Again, this may not be possible for large basis sets, and it will be necessary to recompute some integral batches. It is this potential need to recompute batches of integrals that precludes deduction of a simple formula relating the number of passes over the integrals to the basis set size for the $n_A N^4$ case. However, given the advantages in the vectorization of integral processing in this approach, it is likely that it will out-perform the $n_A^2 N^4$ approach unless pre-screening plays a major role (see below).

It should be noted that - unlike the arrays $\underline{\underline{d}}$ (26) etc. - the arrays of coefficient products $C_{\lambda t}\cdot C_{\sigma u}$ for fixed t, u do not necessarily transform according the totally symmetric irreducible representation of the molecular point group. The process of generating skeleton J and K matrices which are subsequently symmetrized must be treated differently from the SCF case, and a full discussion of the method is given elsewhere [20].

INTEGRAL PRE-SCREENING IN DIRECT MCSCF

Pre-screening is very simple to implement in the construction of $\underline{P}$ (28) etc., it is only necessary to construct a "test density" (9) based on elements of $\underline{d}$, $\underline{e}$ and $\underline{f}$ together. Pre-screening is also easily implemented in the $n_A^2 N^4$ transformation, based on tests of $|C_{\lambda t}\cdot C_{\sigma u}|$ for all tu pairs handled in a given pass. It should be noted that as the basis set increases in size, so that the number of tu pairs which can be handled in a given pass decreases, and the total number of passes increases, the number of small elements in the "test density" cannot increase and may well decrease. This is especially true if the active MOs are localized - this can be done without affecting the final result if a CAS expansion is used [21]. If the test density becomes more sparse the number of integrals to be computed in a given pass will decrease, hence for extended systems with localized active MOs the work required in the course of the $n_A^2 N^4$ approach may well be substantially reduced by pre-screening.

The situation with respect to the $n_A N^4$ approach is not as favourable. Again, tests can be performed on $|C_{\lambda t}\cdot C_{\sigma u}|$, but now all active MO pairs are effectively handled in the one pass, and consequently the test density will be less sparse. Tests could also be made on $|C_{\mu p}\cdot C_{\nu q}|$ for all MOs p, q, and the $\mu\nu$ values of interest for a given array of $[\mu\nu|\lambda\sigma]$, but this also is not likely to be very sparse. Clearly, while the $n_A N^4$ method requires

computation of the integrals only once (or perhaps a few times),
it may well be that almost all of the integrals will need to be
computed in this pass.

CONCLUSIONS

We have discussed aspects of the computation of SCF and MCSCF
wave functions using "direct evaluation" of integrals in
preference to "retrieval from external memory" whenever possible.
Together with the CI case, discussed elsewhere [9], this covers
essentially all of the common LCAO-based approaches to wave
function generation. We have also shown that "direct evaluation"
methods can be formulated to take advantage of vector processors.
While it is not our contention that such methods represent the
only route to wave function generation for large systems and large
basis sets, they seem certain to be of increasing value in the
near future.

REFERENCES

1   J. Almlöf, K. Korsell and K. Faegri jr., J. Comput. Chem.,
     3, 385 (1982).
2   H.P. Lüthi, J.H. Ammeter, J. Almlöf and K. Faegri jr.,
     J. Chem. Phys., 77, 2002 (1982);
     J. Almlöf and K. Faegri jr., J. Amer. Chem. Soc., 105, 965
     (1983).
3   V.R. Saunders and M.F. Guest, Comput. Phys. Commun., 26,
     389 (1982).
4   L.E. McMurchie and E.R. Davidson, J. Comput. Phys., 26, 218
     (1978).
5   D. Hegarty and G. van der Velde, Int. J. Quantum Chem., 23,
     1135 (1983).
6   M. Dupuis, J. Rys and H.F. King, J. Chem. Phys., 65, 111,
     (1976).
7   P. Pulay, Chem. Phys. Lett., 73, 393 (1980); J. Comput. Chem.,
     3, 556 (1982).
8   G.B. Bacskay, Chem. Phys., 61, 385 (1981).
9   P.R. Taylor, to be published.
10   P.D. Dacre, Chem. Phys. Lett., 7, 47 (1970).
11   M. Elder, Int. J. Quantum Chem., 7, 75 (1973).
12   M. Dupuis and H.F. King, Int. J. Quantum Chem., 11, 613 (1977).
13   R.C. Raffenetti, J. Chem. Phys., 58, 4452 (1973).
14   G. Karlström, J. Comput. Chem., 2, 33(1981).
15   E.R. Davidson, J. Chem. Phys., 62, 400 (1975).
16   H.-J. Werner and W. Meyer, J. Chem. Phys., 73, 2342 (1980).
17   P.E.M. Siegbahn, J. Almlöf, A. Heiberg and B.O. Roos, J. Chem.
     Phys., 74, 2384 (1981).
18   P. Jørgensen, J. Olsen and D.L. Yeager, J. Chem. Phys., 75,
     5802 (1981).

19   J.A. Pople, R. Krishnan, H.B. Schlegel and J.S. Binkley,
     Int. J. Quantum Chem. Symp., 13, 225 (1979).
20   P.R. Taylor, Int. J. Quantum Chem., in press.
21   B.O. Roos, P.R. Taylor and P.E.M. Siegbahn, Chem. Phys., 48,
     157 (1980).

COUPLED–CLUSTER METHODS FOR MOLECULAR CALCULATIONS

Rodney J. Bartlett

Quantum Theory Project
University of Florida
Gainesville, Florida 32611

Clifford E. Dykstra

Department of Chemistry
University of Illinois
Urbana, Illinois 61801

Josef Paldus

Department of Applied Mathematics
University of Waterloo
Ontario, Canada N2L 3G1

ABSTRACT

Coupled–cluster (CC) theory for the accurate treatment of electron correlation is presented including its similarities and differences from configuration interaction (CI). Topics addressed include computational aspects of the CC method; extended CC methods that include single, double, and triple excitation operators; and a multi-reference CC technique. Numerical examples illustrate CC results for correlation energies compared to those from full CI and multi-reference CI calculations.

C. E. Dykstra (ed.),
Advanced Theories and Computational Approaches to the Electronic Structure of Molecules, 127–159.
© 1984 by D. Reidel Publishing Company.

INTRODUCTION

One of the purposes of this NATO Advanced Research Workshop
is to examine disparate quantum mechanical methodologies and
clearly delineate the similarities and differences among them.
An equally pertinent question is how these differences impact
the accuracy of the calculation. One of the most promising
newer methods for the accurate treatment of electron correlation
is the coupled-cluster(CC) theory[1-8]. In this article we will
summarize this approach for applications to potential energy
surfaces and other molecular properties, with an emphasis on the
similarities and differences compared to configuration
interaction(CI) methods.

CORRELATED METHODS

Currently, there are three widely used approaches to the
accurate inclusion of electron correlation in molecular applications. These are configuration interaction(CI) (for reviews see
[9,10]), finite order many-body perturbation theory(MBPT)
[11-14], and coupled-cluster(CC) methods[1-8]. They have as a
common objective the improvement of an independent particle
reference function Φ_0 by introducing effects of electron correlation, but they differ in how the approximations are introduced.

For an N-electron system the CI wavefunction may be written
(using intermediate normalization $\langle \Phi_0 | \Psi_{CI} \rangle = 1$) as

$$\Psi_{CI} = \Phi_0 + \hat{C} | \Phi_0 \rangle = \Phi_0 + \sum_n \hat{C}_n | \Phi_0 \rangle \tag{1}$$

where C_n is an excitation operator that introduces n-times
excited configurations with respect to Φ_0 into the CI wavefunction. Labeling the spin orbitals occupied in Φ_0 and their associated creation and annihilation operators by i,j,k, ..., and
the unoccupied spin orbitals and operators by a,b,c, ..., we
can write

$$\hat{C}_n = \sum_{\substack{i<j<k... \\ a<b<c...}} c_{ijk...}^{abc...} (a^\dagger i)(b^\dagger j)(c^\dagger k)..., \quad (n=1,2,... N). \tag{2}$$

The coefficients $C_{ijk\cdots}^{abc\cdots}$ are determined variationally, so that

$$E_{CI} = \langle \Psi_{CI} | H | \Psi_{CI} \rangle / \langle \Psi_{CI} | \Psi_{CI} \rangle = \langle \Phi_0 | H | \Psi_{CI} \rangle . \tag{3}$$

When all excitation levels up to the number of electrons, N, are included, we have the full CI result.

The full CI has several attractive properties. It offers the best possible upper bound to the exact energy in a chosen basis set; it is invariant to any transformation among the orbitals; and for separated molecules A and B it has the size-extensive property that $E_{FCI}(AB) = E_{FCI}(A) + E_{FCI}(B)$[4,8]. When all excitations are not included, these three properties are lost, although any CI calculation is still variational, and a CI calculation that includes all excitations up to a certain excitation level is invariant to separate transformations among the occupied and excited orbitals. Of course, the higher the level of excitations included, the less should be the size-extensivity error. Obviously, the goal of all basis set quantum chemistry is the full CI result, but in view of the difficulties in obtaining this result, alternative routes are sought.

For many years CI calculations restricted to all single and double excitations from a single reference function(CISD) have been routinely applied[15]. The dominant current route in CI is to consider more than one reference function $\{\Phi_\mu, \mu=1,2,\ldots p\}$ and restrict $\hat{C}$ so as to introduce selected categories of excitations like singles and doubles from each of the reference functions[16,17]. This is discussed in other contributions to this volume.

An alternative route to CI is offered by MBPT[8]. There are presently several textbook treatments on this and related subjects[18,19]. Separating the Hamiltonian operator as $H = H_0 + V$, and considering a perturbation expansion in V, the MBPT wavefunction and energy are formally given as

$$|\Psi_{MBPT}\rangle = |\Phi_0\rangle + \sum_{k=1}^{\infty} [Q_0 (E_0 - H_0)^{-1} V]^k |\Phi_0\rangle_L ;$$

$$Q_0 = 1 - |\Phi_0\rangle\langle\Phi_0| , \tag{4}$$

$$E_{MBPT} = \langle\Phi_0 | H | \Psi_{MBPT}\rangle$$

$$= \langle\Phi_0 | H_0 | \Phi_0\rangle + \sum_{k=0}^{\infty} \langle\Phi_0 | V[Q_0(E_0 - H_0)^{-1}V]^k | \Phi_0\rangle_L \tag{5}$$

where the subscript L indicates the restriction to "linked" diagrams[11-13]. The energy is obtained directly from an asymmetric formula and is not variational. The linked-diagram E_{MBPT} energy is size-extensive[8,14,18,19], and this applies to any-order of perturbation theory or even to any linked-diagram approximation regardless of order. If a linked diagram approximation to Ψ_{MBPT} were used as a trial variational function, an energy upper bound would result[20], but the energy would then contain unlinked energy diagrams and therefore not be size-extensive. This emphasizes the general incompatibility of size-extensive and Rayleigh-Ritz variational results until the full CI is reached.

The current state-of-the-art in MBPT is the full fourth-order result, which means all single, double, triple, and quadruple excitation terms through fourth-order relative to a single reference function[21-24]. CI methods cannot usually include all triple and quadruple excitations for realistic applications because of their extremely large number. However, because of the disconnected cluster feature discussed below, both MBPT and CC methods can include the often dominant parts of higher-excitations without the same difficulties that are present in CI. The prime deficiency of MBPT (other than the often practical restriction to a single reference function, which also applies to CC theory and will be discussed later) is the finite-order aspect. Infinite-order methods have advantages in stability and invariance properties. They are also better equipped to handle quasidegeneracies when present[4,25-27].

A most general infinite-order extension of MBPT is offered by coupled-cluster theory[1-8,19b]. Closer inspection of the Ψ_{MBPT} wavefunction shows that all <u>linked</u> wavefunction diagrams are further separated into a <u>connected</u> and a <u>disconnected</u> part[28,29,25,19b] If we define an operator T such that $T \mid \Phi_0 >$ is the sum of all connected parts in all orders, then the exact wavefunction can be expressed as $\exp(T) \mid \Phi_0 >$[28] with the product terms in the expansion of the exponential operator representing the disconnected part. Hence, an alternative to the order-by-order expansion in V of MBPT is to employ the exponential ansatz for the wavefunction,

$$\mid \Psi_{CC} > = \exp(T) \mid \Phi_0>, \tag{6}$$

building in approximations by restricting T to categories of electron cluster operators, T_n. This is the basic idea of the coupled-cluster approach[1-8]. Once $\mid \Psi_{CC} >$ is known, the energy may be obtained from the asymmetric energy formula

$$E_{CC} = <\Phi_0 \mid H \mid \Psi_{CC}> \tag{7}$$

Analogous to the excitation operator $\hat{C}$, the cluster operator T can be decomposed into its one-electron, two-electron, and higher order cluster terms,

$$T = T_1 + T_2 + T_3 + \ldots, \tag{8}$$

with

$$T_n = \sum_{\substack{i<j<k\ldots \\ a<b<c\ldots}} t_{ijk\ldots}^{abc\ldots} (a^\dagger i)(b^\dagger j)(c^\dagger k)\ldots(n=1,2,\ldots N) . \tag{9}$$

The (antisymmetrized) amplitudes $\{t_{ijk\ldots}^{abc\ldots}\}$ of the T_n operator can be related to those of $\hat{C}_n$ in CI. Since necessarily $\Psi_{FCI} = \Psi_{CC}$, we have[30,2-4]

$$C_1 = T_1$$

$$C_2 = T_2 + T_1^2/2 \tag{10}$$

$$C_3 = T_3 + T_1 T_2 + T_1^3/3!$$

$$C_4 = T_4 + T_2^2/2 + (T_1^2/2)T_2 + T_1 T_3 + T_1^4/4! .$$

Eq. (10) illustrates a very important feature of coupled-cluster theory. Namely, categories of CI excitations are further decomposed into clusters and their disconnected products. Since it is far easier to evaluate contributions from products of lower-order cluster operators that sum to n, CC theory offers a way to evaluate most of the contributions of higher-excitations while explicitly considering only relatively low n-values. For the tetraexcited component, Sinanoglu[31] pointed out that $T_2^2/2$ is far more important than T_4 at least for non-metallic systems, and in fact accounts for all quadruple excitations through fourth-order perturbation theory[2-4]. Hence the coupled-cluster doubles(CCD) wavefunction

$$|\Psi_{CCD}\rangle = \exp(T_2)|\Phi_o\rangle, \tag{11}$$

which requires no more amplitudes than a CID calculation, also
offers much of the improvement that would be obtained in a CIDQ
calculation. However, all terms in CCD may be evaluated with
less than m^6 operations for m basis functions, while CIDQ would
require a number of operations proportional to m^8.

Of course, the other primary advantage is that CC methods
are size-extensive, as required by the exponential form of the
wavefunction[30]. Hence,

$$\Psi_{AB} = \exp(T_{AB}) \mid \Phi_{AB} > . \tag{12}$$

When A and B are separated and a localized orbital description
of A and B is used, we have schematically,

$$\exp(T_{AB}) \mid \Phi_{AB} > = \exp(T_A + T_B) \mid \Phi_A \Phi_B >$$
$$= \exp(T_A) \mid \Phi_A > \exp(T_B) \mid \Phi_B > \tag{13}$$

where the second step follows from $\quad T_A \mid \Phi_B > = 0$, assuming
that T_A and T_B commute. The energy expression for separated A
and B is then

$$E_{CC}(AB) = E_{CC}(A) + E_{CC}(B) \tag{14}$$

Note that we assume a localized orbital description for the
wavefunctions of A and B, although the density matrix and energy
expression would be invariant to any similarity transformation.

COUPLED-CLUSTER DOUBLES MODEL

The derivation of the detailed equations of CC theory benefits enormously from employing diagrammatic techniques[2-4].
This is particularly the case when higher than pair clusters are
considered[32-34] or when a symmetry adapted formalism is
required[35]. In the latter case, the time-independent diagrammatic approach[2,3,29,36] can be conveniently combined with
graphical methods of spin algebras[37,38,19b] to obtain either
an orthogonally spin-adapted orbital formalism[32] or, in the
atomic case, the LS (Russell-Saunders) adapted CC formalism[35].
However, for the pair theories the standard wavefunction formalism is quite transparent though slightly more laborious than
the diagrammatic approach[39]. Since one of the purposes of
this article is to emphasize similarities with other methods, we
will initially use the most familiar concepts and consider the
spinorbital version of the simplest realistic model, coupled-
cluster doubles, CCD[4,5], (also called coupled-pair many-
electron theory, CPMET[2] and coupled pair approximation,

CPA[19a]), where T is restricted to only double-excitation terms and Φ_0 is the closed shell SCF solution. Following Ref. [39] we insert the CCD wavefunction, Eq.(11), into the Schrödinger equation, to obtain

$$(H-E)e^{T_2} | \Phi_o \rangle = 0 . \tag{15}$$

Projecting on the left by $\langle \Phi_o |$, we obtain the energy

$$\dot{E}_{CCD} = \langle \Phi_o | H | \Phi_o \rangle + \langle \Phi_o | HT_2 | \Phi_o \rangle = E_{SCF} + \Delta E$$

$$= E_{SCF} + \sum_{\substack{i<j \\ a<b}} \langle ij \| ab \rangle \, t_{ij}^{ab} \tag{16}$$

since H has one and two-electron operators. Notice that this is the same energy expression as in CID [cf. Eq. (3)] but with different amplitudes. The amplitudes $t_{ij}^{ab}(=-t_{ji}^{ab}=-t_{ij}^{ba}=t_{ji}^{ba})$ may be obtained from projection of Eq. (15) onto a sufficient set of doubly-excited configurations, $\{_{ij}^{ab}\}$. Hence

$$\langle _{ij}^{ab} | (H-E_{CCD})e^{T_2} | \Phi_o \rangle = 0$$

$$\langle _{ij}^{ab} | H | \Phi_o \rangle + \langle _{ij}^{ab} | HT_2 | \Phi_o \rangle + 1/2 \langle _{ij}^{ab} | HT_2^2 | \Phi_o \rangle$$

$$- E_{CCD} \langle _{ij}^{ab} | T_2 | \Phi_o \rangle = 0, \tag{17}$$

where E has been replaced by the projected value, E_{CCD}.

Three of the terms in Eq. (17) are the same as would be obtained by CI limited to double exitations. The first term is simply the H-matrix element between a given double-excitation $| _{ij}^{ab} \rangle$ and the SCF reference function $| \Phi_o \rangle$, while the second is the linear combination of H-matrix elements between $| _{ij}^{ab} \rangle$ and all double excitations with coefficients given by the corresponding two-particle amplitudes. The last term picks out a specific amplitude, t_{ij}^{ab}. However, the third term is quadratic in the two-particle amplitudes and therefore differs from CID. In terms of two-electron integrals, it may be shown[2-4,39] that

$$1/2\langle^{ab}_{ij}|\; HT^2_2\;|\Phi_o\rangle = \sum_{\substack{k<\ell \\ c<d}} \langle k\ell\|cd\rangle[t^{ab}_{ij}t^{cd}_{k\ell} + t^{cd}_{ij}t^{ab}_{k\ell} -2(t^{ac}_{ij}t^{bd}_{k\ell}+t^{bd}_{ij}t^{ac}_{k\ell})$$

$$-2(t^{ab}_{ik}t^{cd}_{j\ell} + t^{cd}_{ik}t^{ab}_{j\ell}) + 4(t^{ac}_{ik}t^{bd}_{j\ell} +t^{bd}_{ik}t^{ac}_{j\ell})] \;=\; U^{ab}_{ij} + L^{ab}_{ij}, \quad (18)$$

where U^{ab}_{ij} stands for the unlinked part, the first term in Eq. (18), and L^{ab}_{ij} for the remaining linked parts.

From Eq. (16) we know that

$$U^{ab}_{ij} = \Delta E t^{ab}_{ij} = \sum_{\substack{k<\ell \\ c<d}} \langle k\ell\|cd\rangle\; t^{cd}_{k\ell}t^{ab}_{ij} \qquad\qquad (19)$$

where $\Delta E = E_{CCD} - E_{SCF}$, which cancels the unknown part of $E_{CCD}t^{ab}_{ij}$ leaving the energy independent CCD equations,

$$\langle^{ab}_{ij}|\;H\;|\Phi_o\rangle + \sum_{\substack{k<\ell \\ c<d}} \langle^{ab}_{ij}|\;(H-E_{SCF})\;|^{cd}_{k\ell}\rangle t^{cd}_{k\ell} + L^{ab}_{ij} = 0 \;. \qquad (20)$$

Consequently, with only a slight modification of the CID equations, we obtain a size-extensive linked diagram result that incorporates the most important effects of quadruple excita-tions. We have no more amplitudes to determine than in CID, though we must solve non-linear algebraic equations for the amplitudes instead of an eigenvalue problem. In practice the results obtained from Eq. (20) will be quite close to CIDQ but require far less computation. If L^{ab}_{ij} in Eq. (20) is neglected, one obtains the linearized (L-CCD) version (also called LCPMET[2], D-MBPT(∞)[13b,41], CPA$_O$[19a], and CEPA(0)[40]). Note that this is a size-extensive, linked-diagram approximation and is not CID, the latter being obtained when the amplitudes in the wavefunction $(1+T_2)\;|\Phi_o\rangle$ are determined variationally, thereby satisfying $(T_2 = C_2)$

$$\langle^{ab}_{ij}|\;H\;|\Phi_o\rangle + \sum_{\substack{k<\ell \\ c<d}} \langle^{ab}_{ij}|\;(H-E_{CID})\;|^{cd}_{k\ell}\rangle c^{cd}_{k\ell} = 0 \qquad\qquad (21)$$

This, of course, is an energy-dependent equation. The amplitudes obtained by the iterative solution of Eq. (20) may be inserted into Eq. (16) to obtain the MBPT order-by-order energy diagrams that arise from T_2, giving MBPT results as a by-product of CC calculations.

Since CID neglects quadruple excitations the cancellation expressed in Eq. (19) does not occur, and it is of some interest to assess the relative magnitudes of the U_{ij}^{ab} and L_{ij}^{ab} terms [4,41,43]. These terms first affect the energy in the fourth-order of perturbation theory[4] and may be estimated from considering the iterative solution of the CCD equations. The U_{ij}^{ab} terms eliminate unlinked energy diagrams(U) while the L_{ij}^{ab} terms arise as the corresponding linked quadruple excitation diagrams(L). The fourth-order unlinked diagrams have the value, $E_2 \langle \phi_1 | \phi_1 \rangle$ for E_2, the second-order energy, and ϕ_1, the first-order MBPT wavefunction. U is clearly negative and relatively large, while the linked diagrams are usually a smaller, positive number. For example, for H_2O, U is -17.1mh while L is 3.2mh[41]. Hence, retaining the effects of U through its cancellation in Eq. (19), as is accomplished in L-CCD, is the first important step in a much improved CID calculation[4], in the absence of degeneracies[42]. Because U is negative, however, L-CCD will almost always be a lower bound to CCD[4,41]. (An exception occurs when L is negative, which happens for small basis set calculations[4,25].) L-CCD is obviously always a lower bound to CID[44].

As shown by Bartlett and Shavitt[45], the basis of the widely used Davidson approximation for quadruple excitations[46] is to estimate the effect of the dominant U term from CID parameters and use it to correct CID for quadruple excitations [42,45-48]. In this manner the unlinked diagrams in the CID energy are approximately removed. However, since the L-CCD overshoots the quadruple correction because of neglecting the linked tetraexcited part, some of which (the EPV or "conjoint" parts) actually arises from CI double excitations[41], a more correct estimate of quadruple excitations would fall between L-CCD and CCD. Davidson's formula is $\Delta E_{CID}(1-C_0^2)$, where C_0 is the coefficient of the SCF function in the CI. Derived by Bartlett and Shavitt[45] from MBPT, the renormalized approximation, $\Delta E_{CID} [(1-C_0^2)/C_0^2]$ represents, in fact, the first term of the E_{L-CCD} expansion in terms of CID, as shown by Paldus, et.al.[42,48],

$$E_{L-CCD} = [\sum_{i=0}^{M} C_i/\lambda_i]^{-1}, \tag{22}$$

with λ_i designating the CID energies and C_i the coefficients of $|\Phi_0\rangle$ in the corresponding eigenvectors ($\lambda_0 = E_{CID}$). Since for nondegenerate ground states, $C_0^2 \gg C_i^2$ ($i>0$), the higher terms are negligible and the first term, equivalent to the renormalized Davidson correction, yields practically the L-CCD energy. Similar formulae for the multi-reference case have been proposed by Prime, et. al.[49] and recently by Paldus[48].

An improvement over L-CCD that is still less time-consuming than CCD arises from a study of the L_{ij}^{ab} term in Eq. (18). As written, each of the four non-cancelling terms in L_{ij}^{ab} is about the same magnitude. However, in the orthogonally spin-adapted form[32] of L_{ij}^{ab}, it is found that two of the possible terms, namely those which yield the EPV terms proportional to pair energies, become dominant[50-52]. Restricting the CCD equations to these two terms, which has been called approximate CCD(ACCD)[51,52] or the ACP-D45 approximation[50,53], results in a model that maintains the same invariance properties as CCD and, for many cases, as shown by Dykstra and co-workers[51,52], is very close to the full CCD result. Also, the two dominant terms are the least time-consuming to compute. This should be compared with the CEPA techniques discussed by Ahlrichs[40], Meyer[54], Kutzelnigg[55] and Hurley[19a], and references therein, which do not generally have such invariance properties[52].

The correspondence between CCD and CI has been discussed at length elsewhere[48], but briefly let us consider the CID equations. Designating by D the row of biexcited configurations $\{|_{ij}^{ab}\rangle\}$, so that $|\Psi_{CID}\rangle = |\Phi_0\rangle + |D\rangle c_2$, we define matrices $a = \langle\Phi_0|H|D\rangle$, $b = \langle D|H - E_{SCF}|D\rangle$, with $\Delta E = E_{CID} - E_{SCF}$, and write the CID equations as

$$a\, c_2 = \Delta E , \tag{23a}$$

$$a^\dagger + b\, c_2 = c_2\, \Delta E . \tag{23b}$$

Inserting Eq. (23a) for ΔE into Eq. (23b), we obtain

$$a^\dagger + b\, c_2 - c_2\, a\, c_2 = 0 , \tag{24}$$

which superficially offers a non-linear equation for the CI coefficients. We may compare Eq. (24) with CCD, Eq. (17), which may be rewritten as

$$a^\dagger + b\, t + (1/2)d(t \times t) - t\, a\, t = 0 , \tag{25}$$

where the matrix $\underset{\sim}{d}$ mixes M doubles with M^2 quadruple excitations, and the $\underset{\sim}{t} \times \underset{\sim}{t}$ is a direct product of column dimension M^2. The unlinked term $\underset{\sim}{c_2} a \underset{\sim}{c_2}$ or $\underset{\sim}{t} a \underset{\sim}{t}$ appears in both the CID, Eq. (24) and CCD, Eq. (25). However, in the latter case this term is cancelled by the unlinked term originating in the bi-tetra interaction [cf. Eq. (19)], thus providing the correct size-dependence. Completely neglecting the bi-tetra interaction term, $(1/2)\underset{\sim}{d}$ $(\underset{\sim}{t} \times \underset{\sim}{t})$ results in CID with its size-dependent, unlinked terms.

To go a step further, we could consider the CIDQ equations. We would then have

$$\underset{\sim}{a} \; \underset{\sim}{c_2} = \Delta E \tag{26a}$$

$$\underset{\sim}{a}^\dagger + \underset{\sim}{b} \; \underset{\sim}{c_2} + \underset{\sim}{d} \; \underset{\sim}{c_4} = \underset{\sim}{c_2} \; \Delta E \tag{26b}$$

$$\underset{\sim}{d}^\dagger \; \underset{\sim}{c_2} + \underset{\sim}{q} \; \underset{\sim}{c_4} = \underset{\sim}{c_4} \; \Delta E \; . \tag{26c}$$

The matrix q mixes quadruple excitations with themselves over the $H - E_{SCF}$ operator, and $\underset{\sim}{c_4}$ are the corresponding quadruple excitation coefficients. Whereas Eqs. (24) and (25) only require the determination of M coefficients, Eq. (26) requires approximately M^2 (which is proportional to m^8). Substituting for $\underset{\sim}{c_4}$ in Eq. (26b), Eq. (26c) gives an equation solely in terms of $\underset{\sim}{c_2}$.

$$\underset{\sim}{a}^\dagger + \underset{\sim}{b} \; \underset{\sim}{c_2} + \underset{\sim}{d}(\underset{\sim}{a} \; \underset{\sim}{c_2} \; \underset{\sim}{1} - \underset{\sim}{q})^{-1} \; \underset{\sim}{d}^\dagger \; \underset{\sim}{c_2} - \underset{\sim}{c_2} \; \underset{\sim}{a} \; \underset{\sim}{c_2} = 0 \tag{27}$$

If we approximate the quadruple CI coefficients $\underset{\sim}{c_4}$ by their disconnected part $\underset{\sim}{c_4} \cong (1/2)\underset{\sim}{c_2} \times \underset{\sim}{c_2}$ [cf. Eq. (10)], Eq. (26b) yields again Eq. (25), indicating that CCD offers a superior decoupling procedure to simply setting $\underset{\sim}{c_4} = 0$. Alternatively, Eq. (27) could be used to introduce perturbation corrections to the $\underset{\sim}{c_2}$ coefficients due to quadruple excitation effects, which is similar to the idea behind the A_k and B_k CI methods[56]. By extension, the above development would apply to the remainder of any Hamiltonian matrix, hence the condition that all higher-even fold excitations are decomposable into products of $\underset{\sim}{c_2}$ is contained within Eq. (25).

There are many numerical applications illustrating the significant differences in CCD and CID (in practice single excitations are often included as well)[8]. One is the study of the quartic force field for H_2O[41], another the study of hydrogen

bonded[57] and van der Waals systems[52]. Other ab initio com-
parisons have been made with the full CI for several
examples[22], including some highly degenerate cases[25,26].
Within the PPP and Hubbard model a number of studies comparing
CC to the full CI have addressed extremely difficult problems
involving metallic-like delocalizations[58] and severe
degeneracies[53]. Of course, the size-extensive property for
molecular dissociation is always important[4,59,14,40].

With the recent emergence of several full CI results
[60,61], it would be highly informative if cluster analyses were
carried out in order to offer a numerical assessment of the
validity of coupled-cluster approximations, as has been done for
semi-empirical wavefuctions[62,58,53].

To summarize, in CCD a great deal is gained from higher-
excitation effects from a relatively modest addition to a CI
calculation. After considering computational strategies in the
next section, we discuss extensions of CCD to include T_1 and T_3.

COMPUTATIONAL STRATEGIES FOR CCD

Relative to CID, the additional computational work in CCD
is the evaluation of the L_{ij}^{ab} term in Eq. (18) and the cost of
extra iterative cycles should convergence be slower. With n as
the number of internal or occupied orbitals and m as the total
number of orbitals the number of products of amplitudes used in
Eq. (18) is around $n^4 m^4$ which is just the square of the number
of t_{ij}^{ab} amplitudes. However, by suitable factorization of the
sums in Eq. (18) it is possible to obtain the L_{ij}^{ab} terms for all
ij,ab substitutions with a dependence on n and m of no worse
than $n^3 m^3$[4,5,7]. Since n<m, the dominant part of the calculation (i.e., for large n and m) is the part common to CCD and
CID. This part gives the elements $\langle _{ij}^{ab} | HT_2 | \Phi_o \rangle$ and scales as
$n^2 m^4$. In other words, the cost of computing L_{ij}^{ab} and performing
a CCD calculation instead of just CID is of diminishing relative
importance as m becomes much greater than n.

To see how the computational dependencies in finding L_{ij}^{ab}
arise, we can look at the terms in Eq. (18) either in spin-
orbital form as given, with amplitudes of spin-adapted configurations[7,32], or in coefficient matrix/operator form[52].
The last of these is discussed elsewhere in this volume. All
three achieve the same dependencies on n and m, but implementations differ in the storage and retrieval of intermediate
quantities. The organization for finding the first sum in L_{ij}^{ab}
is representative of what must be done for all the sums.

$$\sum_{\substack{k<\ell \\ c<d}} \langle k\ell \| cd \rangle \; t_{ij}^{cd} t_{k\ell}^{ab} = \sum_{k<\ell} t_{k\ell}^{ab} \sum_{c<d} \langle k\ell \| cd \rangle \; t_{ij}^{cd} \tag{28}$$

$$= \sum_{k<\ell} t_{k\ell}^{ab} \; R_{k\ell ij} \tag{29}$$

The list of intermediate values, R, is n^4 long, and each element
in the list requires $(m-n)^2$ multiplications and additions. The
remaining summation, given in Eq. (29), requires n^2 operations;
to perform this for all the $n^2 m^2$ double substitutions means that
this term scales in cost no worse than $n^4 m^2$. The analogous process in a spin-adapted, matrix-formulated approach[7] is to form
traces of all products of external coefficient matrices with
internal exchange operators. Each trace of a matrix product is
an m^2 operation, and the number of these required is n^4, for a
net dependence of $n^4 m^2$ once again.

Finding the second sum in L_{ij}^{ab} in the spin adapted form is
only an $n^2 m^3$ process. With the neglect of the remaining terms,
one obtains the ACCD[50-53] result. The third sum of L_{ij}^{ab} can
be obtained with only an $n^3 m^2$ cost, so it is the fourth sum that
presents the most serious computational challenge for CCD. At
best, this term has an $n^3 m^3$ dependence, but that can be realized
only with intermediate storage of about $n^2 m^2$ values, the same as
the number of independent amplitudes. In the matrix formulated
approach[7], products of external coefficient matrices and
internal operators are formed and saved at this point. This is
a convenient and effective organization of intermediate values
because each product matrix retrieved is used for computing at
once all the L_{ij}^{ab} terms for the $(m-n)^2$ substitutions from one
pair of internal orbitals. Similar intermediate values, each
labelled by two occupied and two virtual orbitals, are stored in
the implementation of the spin-orbital form by Bartlett,
et.al.[4,25] and Pople et al.[5].

It should be noted, however, that the "spin-orbital form"
is not actually implemented in the naive way that might be
suggested[25]. For example, before programming, the spin-
orbital equations are rewritten and spin factors explicitly
included. The equations are also analyzed to introduce simplifications that result from α and β components having the same
spatial orbital. For such a case only distinct amplitudes are
evaluated. As a result of passively implementing spin-
restricted sums with the same programs as spin-unrestricted
sums, by changing loop limits and inserting appropriate factors

of 2, the work involved in evaluating such equations is essentially the same as using a spin-adapted formulation while
retaining the flexibility of removing spin restrictions for UHF
or other DODS cases. Additional programming considerations for
the spin-orbital CC equations are discussed elsewhere[25].

Most systems of chemical interest are large enough to
require iterative solution of the CCD Eq. (20), just as well as
the CID equation, Eq. (21). This can be done in a number of
ways that achieve very different rates of convergence. Bartlett
and co-workers have routinely used Pade' approximants to enhance
convergence[13b,63]. Purvis and Bartlett have also demonstrated
how convergence can be improved relative to a Jacobi type iteration scheme by using a reduced linear equation method[63]. In
this method, a rectangular transformation using a basis of
amplitudes $\{t^{(1)}\}$ reduces the dimensionality of Eq. (25) which
is then solved directly. The basis is expanded until convergence is achieved. This has its origin in similar methods
for eigenvalue equations proposed by Bartlett and Brandas[64]
and popularized in Davidson's method for eigenvalue problems
[65]. In the reduced linear equation scheme the non-linear,
$t \times t$ term in Eq. (25) can be treated as a linear term
that has an implicit dependence on the amplitudes[63].
Convergence seems best if updated amplitudes are used in
accounting for this dependence. In practice, this method typically achieves convergence to 10^{-6} a.u. in six to ten cycles.
This method has been recognized by Wormer et. al. as a pre-
conditioned conjugate gradient technique[66].

Dykstra and co-workers find that the slow convergence of
the Jacobi type iterative scheme can also be overcome by using
the following perturbative correction at each iteration.

$$\Delta t^{(k+1)} = \frac{a^{\dagger} + bt^{(k)} + 1/2\ d(t^{(k)} \times t^{(k)}) - \Delta E^{(k)}\ t^{(k)}}{a^{\dagger} + bt^{(k)} - \Delta E^{(k)}\ t^{(k)} + \varepsilon\ 1} \qquad (30)$$

$$t^{(k+1)} = t^{(k)} + \Delta t^{(k+1)} \qquad (31)$$

$\Delta E^{(k)}$ is the correlation energy obtained at iteration k, while ε
is the parameter that may be freely adjusted to improve con-
vergence since the numerator in Eq. (30) reaches zero when convergence is achieved. The substitution such as that leading

from Eqs. (17-19) to Eq. (20) which eliminates the explicit
energy dependence in the equation for t is not used since for an
iterative procedure it is not precisely valid until convergence.
Instead, an expectation-like expression [7] can be used for
$\Delta E^{(k)}$, $[\Psi_D = (1 + T_2) \mid \Phi_0 >]$,

$$\Delta E^{(k)} = \langle \psi_D \mid H \mid \Phi_0 \rangle + \langle \psi_D \mid H - E^{(k-1)} \mid \psi_{CCD} \rangle \qquad (32)$$

$$= \underset{\sim}{a} \; \underset{\sim}{t}^{(k)} + \underset{\sim}{t}^{(k)\dagger} \; [\underset{\sim}{a}^\dagger + \underset{\sim}{bt}^{(k)} + 1/2 \; \underset{\sim}{d}(\underset{\sim}{t}^{(k)} \times \underset{\sim}{t}^{(k)}) - \Delta E^{(k-1)} \; \underset{\sim}{t}^{(k)}] \qquad (33)$$

In practice, this straightforward scheme typically yields convergence to 10^{-6} a.u. in five to ten iterations. Common choices
for the parameter, ε, that serves to damp out oscillations in Δt
are values in the range of 0.05 to 0.30 a.u.

A unique aid in obtaining a CC wavefunction is the relation
to the simpler corresponding CI wavefunction. If one considers
the projection energy expressions for CID and CCD, (e.g. Eqs.
(23-25)) along with the Davidson estimate discussed earlier, we
can regard the Davidson estimate as rescaling all the
amplitudes[67],

$$\underset{\sim}{t} \approx (1 - c_0^2)/c_0^2 \; \underset{\sim}{c}_2 \; . \qquad (34)$$

If c_2 or a nearly converged c_2 is known, Eq. (34) or similar
rescalings provide a good initial guess of the CCD amplitudes.
In calculations where this has been tried[67], it was found that
certain calculations requiring eight or nine CCD iterations
could be accomplished by performing four or five CID iterations
to find c_2, rescaling these amplitudes, and then carrying out
four CCD iterations. The savings in this process is that the
computation of the L_{ij}^{ab} terms is avoided for the first half of
the iterations, while the number of iterations is not usually
increased. In practice, this appears to work better than
carrying out L-CCD iterations at first. That, too, will eliminate computation of the L_{ij}^{ab} terms in the first iterations, but
this advantage is easily lost as more iterations tend to be
required because the L-CCD model overshoots CCD.

EXTENDED COUPLED-CLUSTER EQUATIONS

In our discussion of the extended coupled-cluster equations
it is useful to present a slightly more general derivation than

that above.

Defining[36]

$$H = H_o + V, \tag{35}$$

$$H_o \left| \Phi_o \right> = E_o \left| \Phi_o \right> \tag{36}$$

$$H_N = H - \left< \Phi_o \left| H \right| \Phi_o \right> = H_N^{(o)} + V_N, \tag{37}$$

$$H_N^{(o)} = \sum_p f_{pp} \, N[p^\dagger p] \tag{38}$$

$$V_N = f_N + W_N = \sum_{p,q}' f_{pq} \, N[p^\dagger q] + (1/4)\sum_{\substack{p,q \\ r,s}} \left< pq \| rs \right> N[p^\dagger q^\dagger sr] \tag{39}$$

the Schrödinger equation with $\left| \Psi_{CC} \right> = \exp(T) \left| \Phi_o \right>$ takes the form $H_N \exp(T) \left| \Phi_o \right> = \Delta E \exp(T) \left| \Phi_o \right>$, where ΔE would be the correlation energy if Φ_o were the SCF function. The N signifies normal ordering of operators with respect to the h-p vacuum Φ_o, which places all hole creation and particle annihilation operators to the right[36]. Multiplying on the left by $\exp(-T)$, we have the theorem that[32,29,19b].

$$\exp(-T) \, H_N \exp(T) \left| \Phi_o \right> = (H_N \exp(T) \left| \Phi_o \right>)_C = \Delta E \left| \Phi_o \right>, \tag{40}$$

where the C signifies that only connected diagrams are to be retained. Projecting on the left by $\left< \Phi_o \right|$, we find the energy correction to be given by vacuum diagrams,

$$\Delta E = \left< \Phi_o \left| V_N \exp(T) \right| \Phi_o \right>_C \tag{41}$$

in which case, C and L are synonymous. In terms of amplitudes

$$\Delta E = \sum_{\substack{i<j \\ a<b}} \left< ij \| ab \right> (t_{ij}^{ab} + t_i^a t_j^b - t_j^a t_i^b) + \sum_{i,a} f_{ia} t_i^a . \tag{42}$$

The equations determining the n-fold excited amplitudes $t_{ij\cdots}^{ab\cdots}$ are then obtained by projecting Eq. (40) onto the appropriate set of configurations $\left| _{ij}^{ab}\cdots \right>$ obtaining

$$\left< _{ij\cdots}^{ab\cdots} \right| ((H_N^{(o)} + V_N) \exp(T) \left| \Phi_o \right>)_C =$$

$$= D_{ij..}^{ab..} \; t_{ij..}^{ab..} \; + \Lambda_{ij..}^{ab..} \; = 0 \; , \tag{43}$$

where

$$D_{ij..}^{ab..} = f_{aa} + f_{bb} + \ldots - f_{ii} - f_{jj} - \ldots \tag{44}$$

is the Møller-Plesset denominator, and

$$\Lambda_{n,ij..}^{\;\;ab..} = \langle_{ij..}^{ab..} | \; (V_N \; \exp \; (T) \; | \; \Phi_o \; \rangle)_C \tag{45}$$

is given by connected diagrams involving V_N and corresponding T_i components.

For example, for CCD (dropping for simplicity the spin-orbital labels a,b,i,j)

$$\Lambda_2 = \{W_N\} + \{V_N T_2\} + \{V_N T_2^2\}/2 \; , \tag{46}$$

where the symbol $\{....\}_{ij}^{ab}$ designates all non-vanishing connected diagrams with two particle and two hole external lines (labeled by a,b and i,j, respectively) that can be formed by contracting the operator (operator product) enclosed in braces and acting on $| \; \Phi_o \; \rangle$ [36,29].

Considering T_i, i = 1,2,3, we find similarly

$$\Lambda_1 = \{f_N\} + \{V_N T_2\} + \{V_N T_1\} + \{V_N T_3\} + \{V_N T_1 T_2\}$$

$$+ \; \{V_N T_1^2\}/2 + \{V_N T_1^3\}/3! \tag{47a}$$

$$\Lambda_2 = \{W_N\} + \{V_N T_2\} + \{V_N T_1\} + \{V_N T_3\} + \{V_N T_2^2\}/2$$

$$+ \; \{V_N T_1 T_2\} + \{V_N T_1^2\}/2 + \{V_N T_1 T_3\} + \{V_N T_1^2 T_2\}/2$$

$$+ \; \{V_N T_1^3\}/3! + \{V_N T_1^4\}/4! \tag{47b}$$

$$\Lambda_3 = \{V_N T_2\} + \{V_N T_3\} + \{V_N T_2^2\}/2! + \{V_N T_1 T_2\} + \{V_N T_2 T_3\}$$

$$+ \{V_N T_1 T_2^2\}/2 + \{V_N T_1 T_3\} + \{V_N T_1^2 T_2\}/2$$

$$+ \{V_N T_1^2 T_3\}/2 + \{V_N T_1^3 T_2\}/3! \tag{47c}$$

where we neglected coupling to T_4 and T_5 (generally, a two-particle part of V_N permits coupling between T_1 and T_{1+2}). Note that T_4 would start to contribute in the fifth order energy (third order wavefunction) and is usually considered negligible unless a metallic-like delocalization becomes important[58], although for a small molecule Be_2 with a single-reference seems to offer an example of T_4 being non-negligible[68]. Approximating $T \approx T_1 + T_2 + T_3$ will yield all the MBPT energy diagrams up to and including the fourth order in addition to summing disconnected pairs and triples to infinite order.

A second possible approximation is to neglect higher order terms, based on MBPT. Eq. (47) lists individual terms by increasing orders of perturbation theory (considering the most important two-particle part of V_N), and the terms contributing in the same order are arranged according to the number of cluster components involved. With $T_3=0$, Eqs. (47a,b) define the CCSD model[25]. Computationally, this model scales no worse than m^6 for m basis functions, and a general purpose program including all the terms (even quartic T_1 terms) has been written and extensively applied[25,26,22,69]

In the SCF case, $f_N = 0$ and T_1 is at least second-order while T_2 is first-order, hence T_1 should not be too important for SCF cases, and CCD $\approx$ CCSD is a good approximation. Also, a linear treatment of T_1 can often give total energies very close to those obtained using $\exp(T_1)$. However, when other types of orbitals are used, f_N is frequently quite large, and T_1 can even be the dominant part of the result[69]. Furthermore, for such cases, there is no justification for neglecting higher-powers like $\{V_N T_1^4\}$, since certain single-excitation amplitudes may even be greater than unity. As an alternative to evaluating the full set of T_1 and T_2 coupling terms, a transformation to approximate Brueckner orbitals[70,71] will make $T_1 = 0$ for the CCD wavefunction [72]. This orbital transformation uses the amplitudes of the linearly incorporated single substitutions and usually takes one, two or three cycles for T_1 to be negligibly small. However, because the orbital transformation is carried out in the course of a CCD calculation, this procedure can more than double the computational cost.

The T_3 part of Eq. (47) introduces an m^7 asymptotic dependence in the CCSDT equations. However, T_3 must be at least partially included if the energy is to be correct through fourth-order. This has been accomplished in fourth-order MBPT programs[21-24] but not generally in an iterative CC framework. However, Paldus et. al.[33] made such calculations for a minimum basis set BH_3 calculation a decade ago obtaining excellent agreement between the CC and corresponding CI results. Kvasnicka et. al.[73] recently reported similar minimum basis set studies for small diatomics. Now, a general purpose program that includes T_3 iteratively into a CC scheme has been developed by Lee and Bartlett[68]. The $\{V_N T_2\}$ in Eq. (47c) contributes to the second-order wavefunction in W while all other terms are third-order or higher. Similarly the $\{V_N T_1 T_3\}$ term in Eq. (47b) contributes in the sixth-order energy in the SCF case and can be neglected. Hence, by approximating T_3 by its second-order contribution and evaluating the other contributions of T_3 to T_1 and T_2 in Eq. (47a,b), we have an iterative method termed CCSDT-1, that gives the energy correct through fourth-order perturbation theory and the wave-function correct through second-order. The full third-order contribution to T_3 would require that the $\{V_N T_3\}$ term be included, and this would necessitate evaluating the contribution of $\sim m^6$ amplitudes to $\sim m^6$ amplitudes, which is almost prohibitive for realistic applications. However, the contribution of the other two third-order terms could be incorporated fairly easily, but such a partial inclusion of third-order terms would seem less justified. Of course, T_4 would also contribute to the third-order wavefunction, which would introduce $\sim m^8$ step at this level. Just as it was possible to find good approximate CCD models like ACCD, it would be highly desirable to analyze the spin-adapted form of the CCSDT equations to possibly locate similar consistent approximations.

Figure 1 and Table I provide an idea of how well the different CC models compare to full and truncated CI results. These calculations are all relative to RHF reference functions at equilibrium, except for H_2O subject to a symmetric stretch of $1.5R_e$ and $2.0 R_e$[22]. In the latter cases the RHF function offers a very bad reference function that is far from separating correctly for a symmetric displacement. The first two calculations employ DZP basis sets, while H_2O only has a DZ basis. DZP basis sets will give better convergence in CC and MBPT calculations[4], although BH is a poorly convergent case due to the near degeneracy among the Boron 2s and 2p orbitals. The full CI results are from Saxe, et. al.[60] and Harrison and Handy[61].

TABLE I.   Coupled-Cluster Results for the Valence Correlation
           Energy of Molecules Compared to Full CI (a.u.)[a]

MOLECULE	L-CCD	CCD	CCSD	CCSDT-1	Full CI
BH(DZP;Re)	.090194	.085070	.086007	.087222	.087664
HF(DZP;Re)	.197535	.195838	.196824	.199519	.199675
H_2O(DZ;Re)	.145608	.145437	.146240	.147578	.148030
H_2O(DZ;1.5Re)	.208094	.199579	.205402	.209538	.210992
H_2O(DZ;2.0Re)	.330760	.277225	.300733	.315641	.310067

[a]   CCSD results are from Bartlett, Sekino, and Purvis[22],
    CCSDT-1 from Lee, Kucharski, and Bartlett[34] and full CI
    from Harrison and Handy[6]. We appreciate Dr. Hideo Sekino
    performing the L-CCD and CCD calculations to complete this
    table.

Except in the case of H_2O at a symmetric stretch of $2.0R_e$
CCSDT-1 gives essentially 100% of the full CI correlation
energy. In the worst case, it overshoots by almost 2%, no doubt
largely due to the approximations in evaluating CCSDT-1 that
neglects normally positive terms like $\{V_N T_2^2\}$ and, probably,
$\{V_N T_3\}$, that contribute to T_3. CCD is always better than CID
(or CISD), primarily because of its inclusion of quadruple excitation effects. It is apparent from the CI results that CI
quadruple excitations are more important than triple excitations
in every case. However, since CI quadruples account for both
the unlinked and linked effects discussed above, of which the
dominant unlinked part is cancelled from the start in CC theory,
the connected triple excitation terms, T_3, have a propor-
tionally greater effect in the CC method. The inclusion of
triples in CCSDT-1 is quite important in adding the last few
percent of the energy to approach the full CI result. As seen
in Figure 1, even though each of the current calculations uses
an RHF reference function, the average single excitation improvement over CCD for these examples is about equal to that for
triple excitations in CC, being even greater for poorly convergent cases. In other non-HF cases, it can be very large[69].
The best CC results give about the same correlation energy as

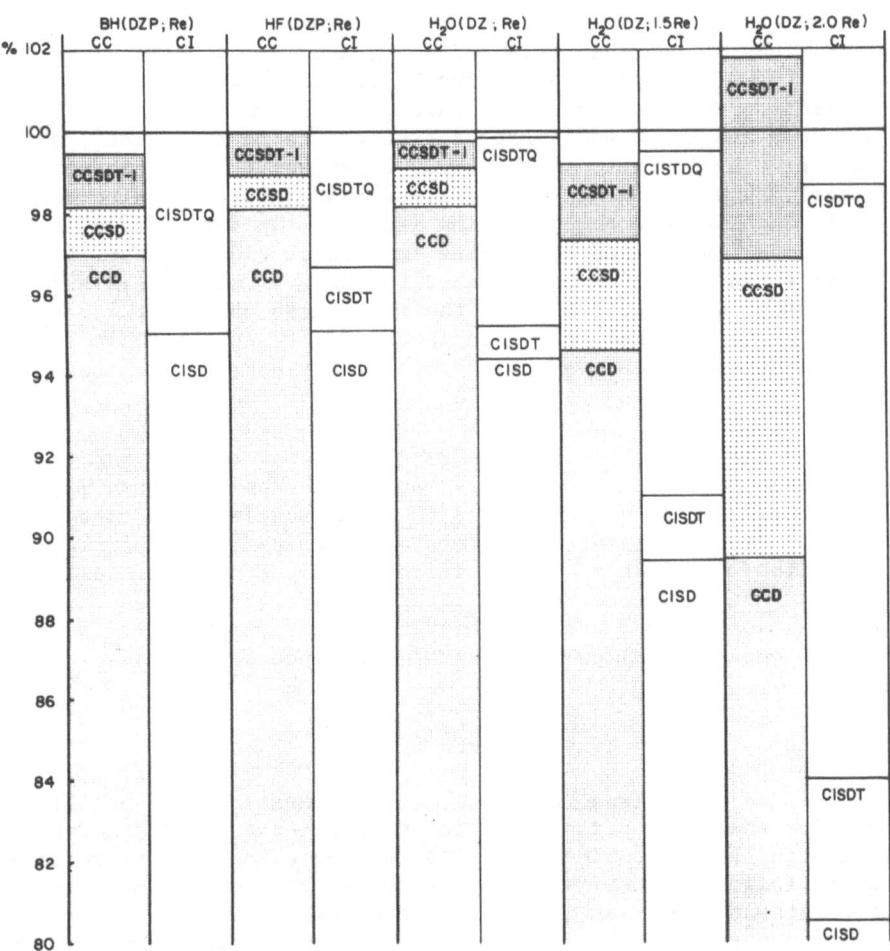

Figure 1.   Comparison of percentages of the correlation
energy obtained by coupled-cluster and configuration
interaction methods.

CISDTQ, although unlike CI, the CC results are size-extensive and do not have to obtain the potentially $\sim m^8$ coefficients as in CISDTQ.

Since the dominant part of CI quadruples is the unlinked part, L-CCD already benefits from this cancellation, only neglecting the smaller linked part. Hence, the difference between CCD and L-CCD would not be expected to be too great[4,5], barring singularities in the L-CCD equations [42]. This is clearly shown in Table I. The linearized theory is quite good until the RHF reference function is poor. In such a case, other configurations are of comparable importance causing near singularities in the L-CCD equations which have a deleterious effect consistent with Eq. (22). In the worst case of H_2O at 2.0 R_e where C_0 for the RHF reference is only 0.764 in the full CI wavefunction, L-CCD overshoots the correct correlation energy by 6.7%.

As the two OH bonds in H_2O are stretched and the RHF reference function becomes very bad, it is apparent from Figure 1 that all the single-reference CC or CI models begin to depreciate. Higher-excitation effects are increasingly important and even CISDTQ or CCSDT-1 cannot introduce important correlating terms. Although some headway might be made by using another orbital set[72,4b], at this point, it seems requisite to modify the CC method to allow for several reference functions.

GENERALIZATION FOR MULTI-REFERENCE SPACES

As is clear from the above, the CC method offers excellent predictions of electron correlation energies and, consequently, potential energy surfaces, if the reference function is an adequate initial approximation. Obviously, the full CI must give the correct basis set limit, so the question is whether the CC method approximates that limit close enough. Even for cases of high degeneracy, single reference CC theory often has enough stability that it will still produce reasonable results [3,4,25,26,27,53]. However, for cases where important correlation effects cannot be introduced into the calculation with less than T_4 for a single reference function (and its concomitant $\sim m^8$ amplitudes) as for H_2O at symmetrically displaced geometries, it is paramount to consider multi-reference methods.

Most multi-reference procedures originate with the finite-order MBPT approach of Brandow[74], where instead of a single Φ_0 we now define a set of important configurations $\{\Phi_\mu\}$ as a reference space. Instead of a single energy, we then get an effective Hamiltonian matrix $\tilde{\bar{H}}$, from which subsequent diagonalization provides several eigenvalues. This adds two degrees of flexibility to the theory: (1) the union of the sets of excitations relative to each $|\Phi_\mu\rangle$ are employed in the calculation; and (2), subsequent diagonalization of $\tilde{\bar{H}}$ takes into account the large coupling among the reference functions completely, rather than depending upon the CC theory to introduce such large terms by summing contributions from the complementary (Q) space, as in the single reference case. Also, the effective Hamiltonian procedure provides a route toward simultaneous evaluation of several eigenvalues.

For the special case of excited or ionized states of systems with a closed shell ground state, Paldus et al.[75] employed the ansatz W exp(T) $|\Phi_0\rangle$, thus exploiting the correlation built into the ground state wavefunction for the excited states, while linearizing the excitation operator, W. Related open-shell, spin-adapted coupled-cluster methods built upon a single reference philosophy have been developed and extensively applied by Nakatsuji and co-workers[76-78] for ionization and excitation energies with excellent results.

More in line with a general multi-reference theory some authors[79,80] attempted a straight-forward generalization of the ansatz of Eq. (6). The problems of such an extension are discussed in Ref.[81]. They can be avoided by a recursive definition involving all multiply ionized states[79,80] resulting in a rather complex formulation. Avoiding the problem of an ill-defined vacuum state for the hole-particle formalism, Lindgren[82] introduced the N-product form of the cluster expansion, $|\Psi_\mu\rangle = N \exp(T)]|\Phi_\mu\rangle$, which enables appropriate disconnected diagram factorization in the open shell case, but exploits a μ-independent cluster operator T. Recently, this ansatz was generalized by Jeziorski and Monkhorst[81] who introduced μ-dependent cluster operators T^μ, one for each reference state $|\Phi_\mu\rangle$. This approach, which is conceptually closest to multireference CI approaches and offers a natural generalization of the single reference theory, is adopted here. The following synopsis follows recent developments by Paldus[48] and Laidig and Bartlett[83].

Consider two idempotent, self-adjoint, and mutually orthogonal projectors,

$$P = \sum_\mu P_\mu = \sum_\mu |\Phi_\mu\rangle\langle\Phi_\mu| \tag{48}$$

$$Q = 1 - P = \sum_r |X_r\rangle\langle X_r| \tag{49}$$

where $\{\Phi_\mu\}$ span the reference space. We will also require that $\{\Phi_\mu\}$ be complete. That is, for any set of orbitals where i,j,k, ... represent the "core" orbitals, m,n,o, ... "active" orbitals, and a,b,c, ... "excited" orbitals, the set $\{\Phi_\mu\}$ contains all excitations within the active space. With these definitions, we can define a wave-operator Ω, such that

$$\Omega \overset{\circ}{\underset{\sim}{\Psi}}_\mu = \underset{\sim}{\Psi}_\mu \qquad (50)$$

where $\{\Psi_\mu\}$ are the exact solutions for state $\{\mu\}$, and $\{\overset{\circ}{\Psi}_\mu\}$ are the corresponding "model" functions,

$$\overset{\circ}{\Psi}_\mu = \sum_\nu \Phi_\nu \, c_{\nu\mu} \, . \qquad (51)$$

It follows that

$$P\Psi_\mu = \overset{\circ}{\Psi}_\mu, \ \Omega P = \Omega, \ P\Omega = P \, . \qquad (52)$$

Inserting Eq. (50) into Schrödinger's equation and using Eq. (52), we have

$$\bar{H}\overset{\circ}{\Psi}_\mu = \overset{\circ}{\Psi}_\mu \, E_\mu \qquad (53)$$

$$\bar{H} = PH\Omega \qquad (54)$$

where the effective Hamiltonian, $\bar{H}$, only operates within the P-space. In matrix form $\bar{H}\underset{\sim}{c}_\mu = \underset{\sim}{c}_\mu E_\mu$ which defines the coefficients in Eq. (51).

Since $\Omega\bar{H}\overset{\circ}{\Psi}_\mu = H\Omega\overset{\circ}{\Psi}_\mu$[82,19b], and trivially, $\Omega\bar{H}Q = H\Omega Q$, we have Bloch's operator equation[84],

$$\Omega H \Omega = H\Omega \qquad (55)$$

which is independent of the energy, E_μ. Bloch's equation provides an approach for obtaining the wave-operator, Ω.

Requiring as much similarity to the single reference CC theory as possible, we choose to define

$$T^\mu = \sum_{i,a} t(a_\mu,i_\mu)\, a_\mu^\dagger i_\mu + \sum_{\substack{i<j \\ a<b}} t(a_\mu,i_\mu;b_\mu,j_\mu)\, a_\mu^\dagger i_\mu b_\mu^\dagger j_\mu + \ldots, \quad (56)$$

where the single particle amplitudes $\{t(a_\mu,i_\mu)\}$ and two-particle antisymmetrized amplitudes $\{t(a_\mu,i_\mu;b_\mu,j_\mu)\}$ etc., are expressed in terms of an orbital partitioning based upon Φ_μ with i_μ,j_μ, ... representing orbitals occupied in Φ_μ and associated operators while a_μ, b_μ ... are the corresponding unoccupied orbitals and operators. Another Φ_λ defines a different partitioning of the orbitals. Of course, any repartitioning can also be interpreted in absolute labeling in terms of the core, active, and excited orbitals introduced above if convenient. This vacuum condition is analogous to that of Hose and Kaldor in MR-MBPT[85]. Due to the complete reference space condition, $T^{\mu\dagger}\,|\,\Phi_\lambda> = 0$ for any μ and λ, while $T^\mu\,|\,\Phi_\mu>$ exclusively creates excitations in the Q-space.

Adopting the ansatz of Jeziorski and Monkhorst[81], that

$$\Omega = \sum_\mu e^{T^\mu} P_\mu \quad\quad (57)$$

and inserting this into Eq. (55), projecting onto Φ_μ, multiplying on the left by e^{-T^μ}, while also using the fact that $<e^{-T^{\mu\dagger}}\Phi_\lambda\,|\, = \,<\Phi_\lambda\,|$ and $Pe^{-T^\mu}\,\bar{H} = \bar{H}$,

$$e^{-T^\mu} He^{T^\mu}\,|\,\Phi_\mu> = \sum_\lambda e^{-T^\mu} e^{T^\lambda}\,|\,\Phi_\lambda><\Phi_\lambda\,|\, e^{-T^\mu} He^{T^\mu}\,|\,\Phi_\mu>$$

$$e^{-T^\mu} He^{T^\mu}\,|\,\Phi_\mu> = \sum_\lambda e^{-T^\mu} e^{T^\lambda}\,|\,\Phi_\lambda>\, \bar{H}_{\lambda\mu}\;. \quad\quad (58)$$

Projection of Eq. (58) onto P defines the matrix elements of the effective Hamiltonian, $\bar{H}$, while projection onto Q provides equations to obtain the T^μ amplitudes. Eq. (58) reduces to the single reference case for one Φ_μ. Since even the single reference L-CCD model can be quite accurate in the absence of degeneracies [41,4,5,42,50,53], and the multi-reference theory should eliminate such singularities, Laidig and Bartlett[83] have recently reported the first ab initio calculations with such a linearized multi-reference coupled-cluster method(MR-LCCM).

To illustrate MR-LCCM, consider Figure 2[83] which plots the difference between the full CI and various other correlated models for the problem previously discussed of H_2O at symmetric displacements. All of the single-reference methods cannot overcome the poorness of the RHF reference function. However, a multi-reference method does not have this problem. Taking the four orbitals in H_2O required for correct separation, $3a_1$, $4a_1$, $1b_2$ and $2b_2$, and using them to define an active orbital set, all possible excitations are constructed. Within this set, an MCSCF calculation is performed (CASSCF)[16]. This defines the reference space and orbitals, from which all single and double excitations give the MR-CISD result. Although the error is ~2mh, it is nearly constant as a function of displacement. The MR-LCCM results are actually closer in an absolute sense to the full CI result, although this is likely to be somewhat fortuitous since linearized methods usually over-estimate the correct correlation energy[4,5,25]. However, in this first attempt the error is not as uniform as the MR-CISD. This may be partially due to two approximations discussed elsewhere[83] introduced for convenience in this initial calculation. Also, perhaps quadratic corrections are significant. Kucharski and Bartlett have discussed the general theory and implemented quadratic extensions[86]. Unlike other multi-reference CC schemes proposed[87], the current method properly includes the important effects of semi-internal excitations. Obviously, as the reference space size is increased, such internal excitations have to be responsible for <u>all</u> the correlation effects. Also, they are well-known to be among the dominant terms in MR-CI calculations, and their neglect is not justified.

Since the foregoing method is size-extensive, it offers a natural vehicle for multi-reference Davidson[46] type approximations. Two possibilities have been suggested by Paldus[48].

To summarize, the proposed multi-reference CC method is an initial attempt. It employs the unitary group treatment for spin-adaptation[88,89]. It is size-extensive. It is comparable in speed to efficient MR-CI techniques and does not require much more time than a single-reference CCSD calculation. As currently implemented it uses a complete reference space, although the theory should be applicable with some modifications and additional restrictions to incomplete reference spaces[85,81]. The latter is an important extension for a generally applicable method and warrants attention. Hence, hope persists that multi-reference coupled-cluster theory will soon take its place beside the single-reference theory to provide an arsenal of powerful tools for molecular structure, properties, and energy surfaces.

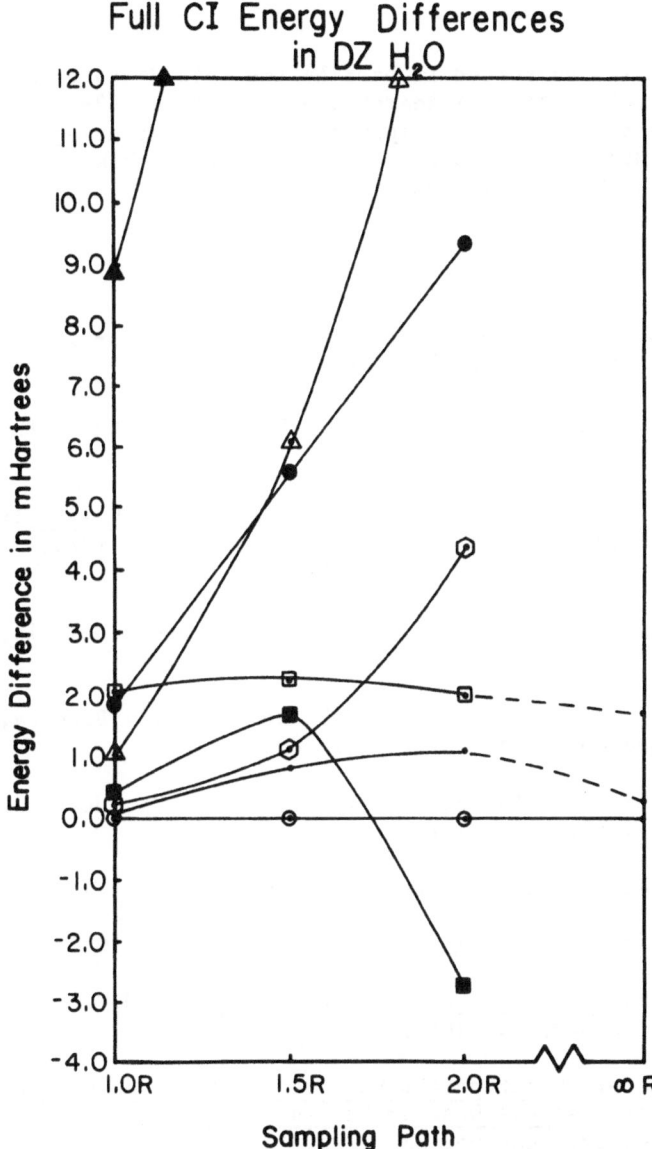

Figure 2.   Comparison of full CI energy differences for various
correlated models along the symmetric O–H stretching
path in DZ H_2O. The various curves are identified
as: (◯) full CI, (▲) CISD, (△) SDTQ-MBPT(4),
(⬡) CCSD, (◯) CISDTQ, (■) CCSD+E_{4T}, (◻) MR–CISD and
(·) MR–LCCM. The dotted line indicates the region
where extrapolation is required since no full CI
results are available.

Acknowledgments

This work was supported in part by the United States Air
Force Office of Scientific Research under grant No. 82-0026 to
the University of Florida. We appreciate the outstanding
assistance of Mrs. Rhonda Montgomery in preparing this
manuscript in a photoready form.

References

1. F. Coester, Nucl. Phys. 1, 421 (1958); F. Coester and H.
    Kümmel, Nucl. Phys. 17, 477 (1960).

2. J. Cizek, J. Chem. Phys. 45, 4256 (1969); Adv. Chem. Phys.
    14, 35 (1969).

3. J. Paldus and J. Cizek, Energy, Structure, and Reactivity,
    ed. D.W. Smith and W.B. McRae, Wiley, New York, 1973; J.
    Cizek and J. Paldus, Phys. Scr. 21, 251 (1980).

4. R.J. Bartlett and G.D. Purvis, Int. J. Quantum Chem. 14,
    561 (1978); Phys. Scr. 21, 255 (1980).

5. J.A. Pople, R. Krishnan, H.B. Schlegel and J.S. Binkley,
    Int. J. Quantum Chem. 14, 545 (1978).

6. P.R. Taylor, G.B. Backsay, A.C. Hurley and N.S. Hush, J.
    Chem. Phys. 69, 1971 (1978).

7. R.A. Chiles and C.E. Dykstra, J. Chem. Phys. 74, 4544 (1981).

8. R.J. Bartlett, Annu. Rev. Phys. Chem. 32, 359 (1981).

9. I. Shavitt, Modern Theoretical Chemistry, ed. H.F.
    Schaefer, Plenum, New York, 1977.

10. H.F. Schaefer, Electronic Structure of Atoms and Molecules:
    A Survey of Rigorous Quantum Mechanical Results,
    Addison-Wesley, Reading, MA, 1972.

11. K.A. Brueckner, Phys. Rev. 97, 1353; 100, 36 (1955); J.
    Goldstone, Proc. R. Soc. London Ser. A239, 267 (1957).

12. H.P. Kelly, Adv. Chem. Phys. 14, 129 (1969).

13. R.J. Bartlett and D.M. Silver, Int. J. Quantum Chem. Symp.
    8, 271 (1974); R.J. Bartlett and I. Shavitt, Chem. Phys.
    Lett. 50, 190 (1977).

14. J.A. Pople, J.S. Binkley and R. Seeger, Int. J. Quantum Chem. Symp. 10, 1 (1976); J.A. Pople, R. Seeger and R. Krishnan, Int. J. Quantum Chem. Symp. 11, 149 (1977).

15. B.O. Roos and P.E.M. Siegbahn, Modern Theoretical Chemistry, vol. 3, ed. H.F. Schaefer, Plenum, New York, 1977.

16. B. Jonsson, B.O. Roos, P.R. Taylor and P.E.M. Siegbahn, J. Chem. Phys. 74, 4566 (1981).

17. H. Lischka, R. Shepard, F.B. Brown, and I. Shavitt, Int. J. Quantum Chem. Symp. 15, 91 (1981).

18. N.H. March, W.H. Young and S. Sampanthar, The Many-Body Problem in Quantum Mechanics, Cambridge University Press, Cambridge, 1967; A.L. Fetter and J.D. Walecka, Quantum Theory of Many-Particle Systems, McGraw Hill, New York, 1971; J. Linderberg and Y. Ohrn, Propagators in Quantum Chemistry, Academic Press, New York, 1973; S. Raimes, Many-Electron Theory, North-Holland, Amsterdam, 1972.

19. A.C. Hurley, Electron Correlation in Small Molecules, Academic Press, New York, 1976; I. Lindgren and J. Morrison, Atomic Many-Body Theory, Springer-Verlag, Berlin, 1982; P. Jørgenson and J. Simons, Second Quantization Based Methods in Quantum Chemistry, Academic Press, New York, 1981.

20. R.J. Bartlett and D.M. Silver, Int. J. Quantum Chem. Symp. 9, 183 (1975).

21. M.J. Frisch, R. Krishnan and J.A. Pople, Chem. Phys. Lett. 75, 66 (1980). R. Krishnan, M.J. Frisch and J.A. Pople, J. Chem. Phys. 72, 4244 (1980).

22. R.J. Bartlett, H. Sekino and G.D. Purvis, Chem. Phys. Lett. 98, 66 (1983).

23. S. Wilson and M.F. Guest, Chem. Phys. Lett. 73, 607 (1980).

24. M. Urban, J. Noga and V. Kello, Theoret. Chim. Acta 62, 549 (1983).

25. G.D. Purvis and R.J. Bartlett, J. Chem. Phys. 76, 7918 (1982).

26. G.D. Purvis, R. Shepard, F.B. Brown and R.J. Bartlett, Int. J. Quantum Chem. 23, 835 (1983).

27. S. Wilson, K. Jankowski and J. Paldus, Int. J. Quantum Chem. 23, 1781 (1983).

28. J. Hubbard, Proc. Roy. Soc. A240, 539 (1957), Proc. Roy.
     Soc. A243, 336 (1958); Proc. Roy. Soc. A199, (1958).

29. J. Paldus, Diagrammatical Methods for Many-Fermion Systems,
     Lecture Notes, University of Nijmegen, 1981.

30. H. Primas, Modern Quantum Chemistry vol. 2, ed. O.
     Sinanoglu, Academic Press, New York, 1965.

31. O. Sinanoglu, J. Chem. Phys. 36, 706 (1962); Advan. Chem.
     Phys. 6, 35 (1964).

32. J. Paldus, J. Chem. Phys. 67, 303 (1977).

33. J. Paldus, J. Cizek and I. Shavitt, Phys. Rev. A5, 50 (1972).

34. Y.S. Lee, S.A. Kucharski and R.J. Bartlett, J. Chem. Phys.
     to be published.

35. B.G. Adams and J. Paldus, Phys. Rev. A20, 1 (1979); A24,
     2302 (1981).

36. J. Paldus and J. Cizek, Adv. Quantum Chem. 9, 105 (1975).

37. E. El Baz and B. Castel, Graphical Methods of Spin
     Algebras, Dekker, New York, 1972.

38. J. Paldus, B.G. Adams and J. Cizek, Int. J. Quantum Chem.
     11, 813 (1977).

39. J. Cizek and J. Paldus, Int. J. Quantum Chem. 5, 359 (1971).

40. R. Ahlrichs, Comput. Physics Commun. 17, 31 (1979).

41. R.J. Bartlett, I. Shavitt and G.D. Purvis, J. Chem. Phys.
     71, 281 (1979).

42. J. Paldus, P.E.S. Wormer, F. Visser and A. van der Avoid,
     J. Chem. Phys. 76, 2458 (1982); Proceedings of Fifth
     Seminar on Computational Methods in Quantum Chemistry, ed.
     W.C. Niewport, Max-Planck Institute für Astrophysik,
     München, 1982.

43. R.J. Bartlett and G.D. Purvis, Annals New York Acad. Sci.
     367, 62 (1981).

44. D. Mukherjee, Chem. Phys. Lett. 79, 559 (1981).

45. R.J. Bartlett and I. Shavitt, Int. J. Quantum Chem. Symp.
     11, 165 (1977); Int. J. Quantum Chem. Symp. 12, 543 (1978).

46. E.R. Davidson, The World of Quantum Chemistry, ed. R. Daudel and B. Pullman, Reidel, Dordrecht, Holland, 1974; S. Langhoff and E.R. Davidson, Int. J. Quantum Chem. 8, 61, (1974); E.R. Davidson and D.W. Silver, Chem. Phys. Lett. 56, 403 (1979).

47. P.E.M. Siegbahn, Chem. Phys. Lett. 55, 386 (1978).

48. J. Paldus, New Horizons in Quantum Chemistry, ed. P.O. Löwdin and B. Pullman, Reidel, Dordrecht, Holland, 1983.

49. S. Prime, C. Rees and M.A. Robb, Mol. Phys. 44, 173 (1981).

50. B.G. Adams, K. Jankowski and J. Paldus, Phys. Rev. A24, 2316 (1981); 2330 (1981).

51. R.A. Chiles and C.E. Dykstra, Chem. Phys. Lett. 80, 69 (1981).

52. S.M. Bachrach, R.A. Chiles and C.E. Dykstra, J. Chem. Phys. 75, 2270 (1981).

53. K. Jankowski and J. Paldus, Int. J. Quantum Chem. 22, 1281 (1982).

54. W. Meyer, Int. J. Quantum Chem. 5, 341 (1971).

55. W. Kutzelnigg, Modern Theoretical Chemistry, vol. 3, ed. H.F. Schaefer, Plenum, New York, 1977.

56. Z. Gershgorn and I. Shavitt, Int. J. Quantum Chem. 2, 751 (1968).

57. M.A. Benzel and C.E. Dykstra, J. Chem. Phys. 78, 4052 (1983).

58. J. Paldus and M.J. Boyle, Int. J. Quantum Chem. 22, 1281 (1982).

59. L.T. Redmon, G.D. Purvis and R.J. Bartlett, J. Am. Chem. Soc. 101, 2856 (1979); G.F. Adams, G.D. Bent, G.D. Purvis and R.J. Bartlett, Chem. Phys. Lett. 81, 461 (1981).

60. P. Saxe, H.F. Schaefer and N.C. Handy, Chem. Phys. Lett. 79, 202 (1981).

61. R.J. Harrison and N.C. Handy, Chem. Phys. Lett. 95, 386 (1983).

62. J. Cizek, J. Paldus and L. Sroubkova, Int. J. Quantum Chem. 3, 149 (1969).

63. G.D. Purvis and R.J. Bartlett, J. Chem. Phys. $\underline{75}$, 1284 (1981).

64. R.J. Bartlett and E.J. Brandas, J. Chem. Phys. $\underline{59}$, 2032 (1973); $\underline{56}$, 5467 (1972).

65. E.R. Davidson, J. Comput. Phys. $\underline{17}$, 87 (1975).

66. P.E.S. Wormer, F. Visser and J. Paldus, J. Comput. Phys. $\underline{48}$, 23 (1982).

67. C.E. Dykstra, Chem. Phys. Lett. $\underline{88}$, 202 (1982).

68. Y.S. Lee and R.J. Bartlett, J. Chem. Phys., in press.

69. W.D. Laidig, G.D. Purvis and R.J. Bartlett, Int. J. Quantum Chem. Symp. $\underline{16}$, 561 (1982); Chem. Phys. Lett. $\underline{97}$, 209 (1983).

70. K.A. Brueckner, Phys. Rev. $\underline{96}$, 508 (1954); R.K. Nesbet, Phys. Rev. $\underline{109}$, 1632 (1958).

71. P.O. Löwdin, J. Math. Phys. $\underline{3}$, 1171 (1962).

72. P.G. Jasien and C.E. Dykstra, Int. J. Quantum Chem. Symp. $\underline{17}$, 289 (1983).

73. V. Kvasnicka, Phys. Rev. A$\underline{25}$, 671 (1982); V. Kvasnicka, V. Laurinc and S. Biskupic, Phys. Reports $\underline{90}$, 159 (1982).

74. B.H. Brandow, Rev. Mod. Phys. $\underline{39}$, 771 (1967); Adv. Quantum Chem. $\underline{10}$, 187 (1977).

75. J. Paldus, J. Cizek, M. Saute and A. Laforgue, Phys. Rev. A$\underline{17}$, 805 (1978); M. Saute, J. Paldus and J. Cizek, Int. J. Quantum Chem. $\underline{15}$, 463 (1979).

76. H. Nakatsuji and K. Hirao, J. Chem. Phys. $\underline{68}$, 2053 (1978); $\underline{68}$, 4279 (1978).

77. H. Nakatsuji and T. Yonezawa, Chem. Phys. Lett. $\underline{87}$, 426 (1982).

78. H. Nakatsuji, K. Ohta and K. Hirao, Int. J. Quantum Chem. $\underline{20}$, 1301 (1981).

79. D. Mukherjee, R.K. Moitra and A. Mukhopadhyey, Mol. Phys. $\underline{30}$, 1861 (1975); $\underline{33}$, 955 (1977); Pramana $\underline{4}$, 247 (1975).

80. F. Coester, Lectures in Theoretical Physics, ed. K.T. Mahanthappa and W.E. Brittin, Gordon and Breach, New York, vol 11B, 1969.

81. B. Jeziorski and H. Monkhorst, Phys. Rev. A. 24, 1668 (1981).

82. I. Lindgren, Int. J. Quantum Chem. 12, 33 (1978).

83. W.D. Laidig and R.J. Bartlett, Chem. Phys. Lett., in press.

84. C. Bloch, Nucl. Phys. 6, 329 (1958).

85. G. Hose and U. Kaldor, J. Phys. B. 12, 3827 (1979).

86. S.A. Kucharski and R.J. Bartlett, to be published.

87. A. Banerjee and J. Simons, Int. J. Quantum Chem. 19, 207 (1981); J. Chem. Phys. 76, 4548 (1982).

88. J. Paldus, J. Chem. Phys. 61, 5321 (1974).

89. I. Shavitt, Int. J. Quantum Chem. Symp. 11, 131 (1977); Int. J. Quantum Chem. Symp. 12, 5 (1978).

STATE-SPECIFIC THEORY OF ELECTRON CORRELATION IN EXCITED STATES

Cleanthes A. Nicolaides

Theoretical and Physical Chemistry Institute
National Hellenic Research Foundation
Vas. Constantinou Avenue 48
Athens 501/1, Greece

An approach to the many-electron problem is reviewed which, apart from being practical and general, essentially addresses the "basis set problem". It puts emphasis on the separate and appropriate choice and optimization of the function spaces describing the zeroth order and the remaining part of the N-electron wavefunction, especially in excited states. The calculated wavefunctions are compact and physically transparent while the results for properties which are obtained from them are reliable. The characteristic numerical examples which are given cover a wide spectrum: the importance of higher than pair subshell correlation clusters in certain excited states, (triple excitations in $Cl \ KL3s3p^6 \ ^2S$), positions of resonances and autoionizing states in He^- and He, the "sudden polarization effect" in ethylene, the transition probabilities to the valence-Rydberg ^2S series of boron, the second-order perturbation theory calculation of electron correlation in CH_4, the 1s binding energies in atomic as well as in metallic Be, and the Hartree-Fock calculation of transition probabilities to multiply excited states and of the binding of the first excited state of the noble gas dihydrides, HeH_2, NeH_2 and ArH_2.

I.   INTRODUCTION

Advances in the quantum theory of electronic matter and of physico-chemical processes have been made possible by a combination of physical intuition and related models (e.g. the independent particle model, IPM), efficient analysis of and practical approaches to the many-body problem, and the revolution in

C. E. Dykstra (ed.),
Advanced Theories and Computational Approaches to the Electronic Structure of Molecules, 161–184.
© 1984 by D. Reidel Publishing Company.

computational capabilities via the use of computers. The theoretical foundations of these advances lie in the possibility of
transforming the multi-dimensional Schrodinger differential equation into a matrix equation, through the use of one-electron or
many-electron function spaces. The choice, manipulation, optimization, application and interpretation of these function spaces
are the essential components of progress in the study of electronic matter.

In this short article I review an approach to the many-
electron problem which emphasizes the importance of the proper
choice and optimization of zeroth order and virtual function
spaces, especially in excited states, depending on the property
under examination. This approach has been applied over the years
to a large variety of systems and transition processes. The few
numerical examples presented here are consequences of the concepts and analyses characterizing it.

Even though the total energies of ground states have always
served as testing grounds for advanced many-body methods, from
the chemical or spectroscopic point of view it is energy differences and probabilities for transitions which are of interest.
In this respect, ground state energy computations are worth careful attention and high accuracy only in special cases such as
questions of detail of structural stability and of closely lying
states of different symmetry or certain cases of sensitive vibrational structure. Conceptually, the computational problem is
related mainly to the number of electrons and to possible com-
plexities in the dissociation region. Hence, most ground state
many-body theories are, by now, algorithm oriented. On the other
hand, excited states below or above the ionization threshold
present additional formal and conceptual difficulties and peculiarities which must be considered whenever there is need for a
rigorous treatment. In order to put the approach of this article
in a suitable perspective, in the next section I recall some of
these peculiarities.

II.   PECULARITIES AND DIFFICULTIES PERTAINING TO EXCITED STATES

Current advances in experimental techniques and observations
made possible by sources such as sychrotron radiation, lasers,
and high quality particle beams and by high resolution spectro-
scopies, allow the detection and study of a variety of excited
states in the discrete or the continuous spectrum. As compared
with that of ground states, especially when these are in equilibrium geometries, their quantitative analysis includes additional
important difficulties.

In general, an open-shell, multiconfigurational zeroth
order description is needed. Thus, the computational many-electron theories must be accordingly practical and consistent.
Furthermore, given the increased possibility of strong near-
degeneracy effects (almost certain to occur as a function of
geometry), these theories must ensure convergence of their solutions to the states of interest, or alternatively, must be sufficiently flexible to be able to identify and characterize their
solutions correctly.

Many excited states are involved in a variety of photo-
physical and photochemical processes of increased complexity.
From published results and from computational experience the
conclusion that can be drawn, in general, is that it is not necessary to obtain very accurate eigenvalues of the Schrodinger
equation in order to have wavefunctions which can yield accurate
(say within 2%-15% error) diagonal or off-diagonal properties.
Inversely, even when an accurate energy solution is obtained, it
does not follow that all other properties can be calculated
accurately in all cases. Examples are the hyperfine structure,
dipole or transition moments of excited states exhibiting strong
near-degeneracy effects, radiationless transition rates etc.
These facts are related to choices of function spaces. Thus,
there is a need for proper characterization of the wavefunctions
and for a relatively consistent analysis of the role of electron
correlation on phenomena such as autoionization, photochemical
reaction paths and cross-sections, fine and hyperfine structure,
presence of external fields, multiphoton transition probabilities
etc., which allow accurate explanations and reliable predictions
without necessarily requiring exorbitant computational effort.

For highly excited states, the function space corresponding
to the continuous spectrum acquires physical significance. Thus,
contrary to the case of the ground or bound excited states where
the continuous spectrum is represented by closed channels only,
whose contribution to electron correlation can be computed by
ordinary variation of perturbation theories, now the correct treatment of electron correlation demands the explicit consideration
of open channels as well. In general, the function spaces and
methods of optimization for open channels are different from
those applicable to closed channels only.

Given the increasing density of states as a function of
upgoing energy, small perturbations can in principle have a
significant effect on the character of the wavefunctions and
related properties. Thus, relativistic corrections could become
non-negligible even in cases where their off-diagonal matrix
elements are small.

Configuration-interaction as well as perturbation theory

considerations indicate that the relative significance of correlation clusters higher than pair may be larger in excited states and may vary as a function of energy. The existence of this "energy dimension" (as distinguished from the well-known "number of electrons(N) dimension") implies that the results of the "singles and doubles" algorithm may not be as consistent in the case of highly excited states, even for small N systems. This idea was put forward a few years ago (1). In this article I present the first numerical evidence of the "energy dimension" from an application to the $T_{2p^3}(r_1, r_2, r_3)$ triplet clusters in $C\ell(KL\ 3s^23p^5\ ^2P^0)$ and in $C\ell(KL\ 3s3p^6\ ^2S)$. I point out that the dependence on energy need not be monotonic since the types of states and of electron-electron interactions play a role as well. Finally, another possibility, the complete breakdown of the Born-Oppenheimer approximation in highly excited states, is not addressed here.

III. CHOICE AND OPTIMIZATION OF FUNCTION SPACES FOR THE CALCULATION OF EXCITED STATES OF THE DISCRETE OR THE CONTINUOUS SPECTRUM

In this section, only the basic ingredients of the theory are presented. More detailed presentations and full lists of references can be found in recent review articles (1,2) and in other publications (3-8).

The usual approach to the calculation of the effects of electron correlation on properties of atomic or molecular excited states is to use a single set of one-electron basis functions. This set is either found in the literature or is obtained with respect to a single reference function which is considered to be appropriate. The various applications of configuration-interaction theory (9,10), many body perturbation theory (11,12), Green's function and electron propagator theory (13,15), close-coupling and R-matrix theory (16,17), to properties of excited states in the discrete or the continuous spectrum have used such orthonormal basis sets. Obviously, one expects that by enlarging these sets the reliability of the results increases. However, it is still not clear how to enlarge them economically to computationally meaningful sizes, even with today's super computers. The n^4-n^6 dependence of the overall calculation (where n is the number of basis functions) imposes formidable difficulties even for small systems, especially for scattering type calculations of resonances. Furthermore, and most important, even large basis sets which are chosen arbitrarily may not yield good descriptions of excited states, especially when there is heavy mixing among correlation vectors and configurations representing closed or open channels. This is so mainly because the orbitals entering in the single or multiconfigurational zeroth order representation of the excited state are not necessarily optimized for that state and their inadequacy is not made up by CI. This fact can be observed easily in atomic structure calculations where the excited state orbitals corresponding to a particular configuration can be computed accurately by a variety of methods, e.g. hydrogenic, two STO SCF, configurational-average Hartree-Fock, Restricted H-F, Multiconfigurational HF. Take, e.g., the nsnp first excited configurations of the alkaline earths. A HF np for the $^1p^0$ state looks very different from the HF np for the $^3p^0$ state and both look different from a HF np from the np^2 ^1S configuration. Obviously then, the accuracy, convergence and interpretation of a many-electron calculation of an excited state property could depend crucially on the choice of the zeroth order orbitals. Observations and conclusions of this type for molecular structure calculations of excited state properties can be found in Refs. 18-20. Nitzsche and Davidson (18) point out the inadequacy of using ground state or triplet ππ* orbitals to describe the singlet ππ* state of amides, and Brooks and Schaefer (19) and Bonacic-Koutecky et al (20) describe the sensitivity of the sudden polarization effect to basis sets, in spite of the magnitude of their calculations.

The theory of excited states which is outlined below is state-specific and it refers to both bound and autoionizing states. By state-specific I mean that the choice and optimization of the one-electron basis sets and many electron configurations and correlation vectors correspond to the orbital structure of the system for each geometry and each energy region of interest (roughly).

Since I report on work which has been carried over the years, I feel it is appropriate to give specific credit to the people involved in it in the text. Aspects of it, with applications to highly excited atomic states (21-23), were started in 1970 in collaboration with D. R. Beck. The aim was the computation of those one and two spin-orbital correlation effects which have been classified by Sinanoglu (24,25) as "nondynamical correlation". In 1972, a subshell (rather than spin orbital) cluster expansion formalism was adopted in order to obtain a one to one correspondence between spectroscopy and the wavefunction expansion. Term by term, the subshell expansion does not have a one to one correspondence with the spin-orbital expansion. (e.g. subshell singles may contain spin-orbital pairs etc.) The aim was to develop efficient procedures for obtaining either selected few or all the one- and two-subshell correlations by minimizing corresponding energy functionals and diagonalizing small CI matrices. The publication of Froese-Fischer's 1972 numerical MCHF computer code for atoms (26) opened the way for obtaining essentially exact solutions of the zeroth order multiconfigurational space which was present in the theory and was called the "Fermi-Sea" (1,27), and of the Rydberg and valence configurations present in problems of valence-Rydberg interactions (1,2). Analysis of the contribution of the

various subshell correlation effects to inner electron spectroscopy, allowed the identification of specific effects throughout
the periodic table with exceptional importance (27). In 1975, a
many-body theory of photoabsorption for systems with high symmetry was published (2,28) which reduces the magnitude of the
problem by computing only selected single and pair subshell correlations. An application to a molecular system, H_2O, was carried out in collaboration with G. Theedorakopoulos (6). Rigorous
extensions to quasibound states of the continuous spectrum have
been worked out in collaboration with Y. Komninos and D. R. Beck
either via the transformations to complex coordinates and the
complex energy plane (5) or via the application of reaction-
matrix (K-matrix) scattering theory (3). Furthermore, in collaboration with G. Aspromallis and D. R. Beck, a combination of discrete and scattering function spaces in the presence of the
relativistic Breit-Pauli Hamiltonian has been achieved, for the
treatment of electron correlation and relativistic effects in the
continuous spectrum of atoms (29). Finally, two basic elements
of this approach have been incorporated systematically in recent
molecular applications (see next section) by Beck (30a), Beck and
Kunz (30b) and Petsalakis et al (7). Separate optimization of the
virtual orbital space in terms of atomic, Gaussian orbitals (1)
has yielded good convergence in H_2 (30a) and CH_4 (30b) with small
basis sets. Similarly, a very small 6 x 6 CI calculation with
allowance for nonorthonormality among the separately obtained excited configurations has yielded convergence for the "sudden
polarization effect" in excited ethylene (7).

For an arbitrary Born-Oppenheimer excited state, the N-
electron wavefunction, Ψ, can be written symbolically in a subshell (symmetry adapted) cluster expansion:

$$\Psi(E) = \phi_{FS} + \phi_{FS}^{-1}\,\sigma + \phi_{FS}^{-2}\,\pi + \phi_{FS}^{-3}\,\tau + \ldots \tag{1}$$

An alternative form, most often met in ground state expansions, is based on the Hartree-Fock zeroth order vector. I.e.
the multiconfigurational ϕ_{FS} (see below) is replaced by ϕ_{HF}. As
is well known, in this case, the relative importance of the
various correlation clusters, σ, π, τ, etc. changes.

$\Psi(E)$: The energy dependence refers only to the case where
the state is in the continuum and therefore the wavefunction has,
rigorously, scattering components corresponding to the open
channels.

ϕ_{FS} = The multiconfigurational Fermi-Sea vector (1,27) whose
orbitals are optimized self-consistently for the root corresponding to the state of the interest. Near-degeneracy cases are

difficult to converge. If a suitably convergent MCHF code is
not available, separate SCF calculations for the important configurations and subsequent diagonalization taking into account
possible nonorthonormalities, yield an equally good zeroth order
description--at least in the atomic cases which we have tested.
In the case of atoms, simple techniques have allowed convergence
of numerical MCHF even in the case of diffuse resonances (3).

σ, π, τ ... symbolize the single, pair, triple ... subshell
excitations into the virtual function space which is orthogonal
to that of the occupied, Fermi-Sea space. They are divided into
two parts: the localized and the asymptotic. The first corresponds to the function space describing the closed channels.
The second corresponds to the space describing the open channels.
The σ have, in most cases, only closed channel components. Excep-
tions are cases of shape resonances (31). The most important
electron-electron interactions in small N systems are those contained in π. As is well known, they contribute to the energy
the most. Furthermore, in the case of autoionizing states it is
the asymptotic part of the pair containing the open channel that
is responsible for the mechanism of autoionization--to the degree that the pairs are decoupled (5,21).

As it will be seen below with the numerical examples, the
choice of a HF or a MCHF for the excited state of interest as a
zeroth order description improves the qualitative and quantitative analysis of the many-body problem for excited states considerably. For one thing, it allows possibilities for judicious
and systematic analyses of correlation functions and energies
transferabilities. Also, it allows the deduction of fairly
accurate information for certain properties and processes, such
as multielectron transition probabilities to doubly excited
states (8), without much computational effort. Furthermore, it
forms the essential basis for the calculation of compact and
accurate correlated excited state wavefunctions (1,3). This is
done in the following manner. Given the orbital structure of the
zeroth order vector, pair and higher correlation functions can be
divided into one part containing at least one Fermi-Sea orbital
(which is not fully occupied in the main configuration) and another containing none (i.e. only virtual orbitals). For example:

$$\pi^{\alpha} = \pi^{\alpha}_{FS-v} + \pi^{\alpha}_{v} \tag{2}$$

where α is the index for the subshell pair being substituted by
π^{α}.

The strategy for solving for π depends on the relative position of the excited state in the spectrum (i.e. whether it is

the lowest, embedded in the discrete or embedded in the continuous spectrum). Note that this position often depends on geometry or nuclear charge, both of which can alter the sequence of states and the character of the spectra. (For example, the H_2^- $^2\Sigma_u^-$ state changes from a resonance to a bound state as a function of internuclear distance. The well-known near-degeneracy configuration $1s^2 2p^2$ ^1S of the Be sequence, corresponds to a state in the continuum for $Z \leq 4$, with an infinity of lower states of the same symmetry, and to the second discrete state for $Z \geq 5$).

A.   Excited State Lowest of its Symmetry

Each π^α is expanded in terms of a few STOs or GTOs whose exponents are initialized appropriately (1) and are then optimized either by minimizing the second order energy or the energy of the diagonalized pair matrix. The N-electron function is put together including the σ--which are also expanded in terms of STOs which are optimized from the total CI. Further optimization of the πs from the total CI is not necessary, although computationally feasible. The resulting nonorthonormality among the correlation functions can be taken into account by explicit computation of N-electron integrals. In the case of atoms, configurational symmetry reduces the number of such integrals significantly. In this approach, good convergence is obtained from small matrices (1,32). A discussion and results related to the possibility of using optimized atomic virtual Gaussian orbitals for use in molecular studies is given in Ref. 1. Here, I note that the explicit consideration of nonorthonormal pairs for molecular ground states was introduced by Meyer in his theory of self-consistent electron pairs (15).

B.   Excited State Embedded in the Discrete Spectrum

This state can have either valence or Rydberg or a mixed such character. The last situation occurs when there are near degeneracies between a valence configuration and a Rydberg series in the presence or not of another valence configuration belonging to the Fermi-Sea. This type of interaction corresponds to the π_{FS-v} term. In this case, all terms of π_{FS-v} near the strong mixing region between the valence configurations and the Rydberg series, are represented by HF solutions. Nonorthonormality may again become important. The remaining of π_{FS-v} and π_v is expanded and optimized as before. The final solution is obtained from a total CI.

In this context, I note that low lying excited singlets of molecules can be treated efficiently by optimizing the corresponding configurations separately. This expectation has led us (7) to the development of a CI approach which incorporates NON according to the general method of King et al (33). A 6 x 6

NON-CI yields good results for the sudden polarization effect
in ethylene (7.19,20) (see below).

C.   Excited State Embedded in the Continuous Spectrum

The continuous spectrum has traditionally been outside the
realm of quantum chemistry. Yet, as the need for the accurate
study of highly excited states increases, the link between many-
electron and scattering theory and computational approaches is
strengthened. For atoms, suitable scattering function spaces
(in numerical form) are possible due to their spherical symmetry.
For molecules, single center expansions in conjunction with
linear algebraic equations (34) are, of course, slow converging
but are suitable for vector computers. Multicenter analytic
$\mathcal{L}^2$ methods are also being developed and studied (35). Nevertheless, in spite of the considerable progress, the consistent, reliable and practical representation of the continuous spectrum
in molecules is still a desideratum. The short presentation which
follows refers to the atomic scattering function space and the
quasi-bound, autoionizing states.

In the case of an isolated autoionizing state the calculation
can proceed as follows (3,4): First, the correlated square-inte-
grable part is obtained by the methods described above. The
asymptotic pair function corresponding to the open decay channel
is then expanded in terms of a Fermi-Sea bound orbital and a
fixed core scattering function of the form $u_{E1} \underset{\tau \to \infty}{\tilde{\sim}} \sin(K\tau+D_E)$,
where the phase shift D_E is related to the effective continuum-
bound function space interaction, which is given by the K-matrix
as $\tan D_E = -\pi K(E)$.

In this independent asymptotic-pair calculation, the contribution to the energy shift is additive, as is the case for the
first order pair calculation of the bound state problem. I note
that instead of real scattering functions, complex coordinates
can be introduced and the asymptotic pairs can be calculated from
novel variational-iterative procedures (5,36). The calculation
of the asymptotic pair-pair interactions, in the presence of the
other pairs, in arbitrary many-electron systems, is a complex
computational problem although the formalism for it exists.

IV.   APPLICATIONS

The conceptual as well as the computational point of view
outlined above is that the many-electron treatment of excited
states and their properties requires flexible and general theories of electronic structure which become more economic and
physically more transparent if they are state-specific. Attention to state specificity and related analysis may result into

small but accurate "property-specific" wavefunction without
necessarily going through the effort of obtaining very accurate
total energies, except when necessary. The numerical examples
presented below for a variety of properties are in this context.

A.   Multielectron Correlation Effects in Excited States: The
      Energy Dimension

Since the early 1960's (25) it is known that calculations
on ground states which take into account "single and double"
spin-orbital substitutions yield most of the correlation energy
for small N systems. "Quadruples" formed from "unlinked pairs"
(24) are next in importance. The contribution from "triples"
is negligible. However, as it was pointed out in Ref. (1) and
Section II, in excited states the "energy dimension" may change
the relative contributions.

In order to demonstrate this theoretical prediction I have
chosen $C\ell$ in its ground state, KL $3s^2 3p^5 \ ^2P^0$, and its valence
excited state, KL $3s3p^6 \ ^2S$. In collaboration with G. Aspromallis,
electron correlation calculations in the 7-electron valence space
were carried out in order to obtain the magnitude of the τ_{3p}^3
triple cluster contribution to the total energy in the ground
and in the excited state. The hole-filling τ_{3p}^3 for the ^2S state
contributes significantly to the total energy via the Π_{3p}^2,
$3s3p^6 \leftrightarrow 3s^2 3p^4 3d, \ 3s^2 3p^4 4s$. The results are shown in Table 1 and
show the predicted significant increase in the relative impor-
tance of the triple excitations in the excited states.

B.   Choice of Basis Sets and the Characteristics of Excited
      State Wavefunctions

The transitions in boron $1s^2 2s^2 2p \ ^2P^0 \rightarrow 1s^2 2s^2 ns, \ 1s^2 2s2p^2 \ ^2S$
offer a good example of the sensitivity of properties such as
transition probabilities to the choice of basis sets in the case
of valence-Rydberg interactions (2,37). Table 2 shows three
columns with theoretical values for the oscillator strengths in
the length formulation (2) and one with the available experimental
ones. It is seen that there is a large difference between an
extensive calculation which used a common basis set (38) and our
small one which used state-specific excited HF functions and only
part of the additional correlation (37). Not only the individual
transitions but the whole oscillator strength distribution is
different. An indication of the level of accuracy of our results
is obtained by comparing them to the two available experimental
values (39,40) as given in Table 2.

In 90° - twisted ethylene, there are two zwitterionic singlet
states, the Z and the V(41), which, in C_s symmetry, are described
in zeroth order by

Table 1. The "Energy Dimension" in Excited States

Cℓ KL 3s3p^6 ^{2}S	Total Energy (a.u.)	Energy Lowering (a.u.)
Hartree Fock	-458.914549	0
Fermi Sea, MC	-459.13227056	-0.21772
CISD from HF (95 configurations)	-459.17928296	-0.0470
CISD with only the hole filling triples 3p^3 → 3sv$_1$v$_1$' (130 configurations)	-459.18768259	-0.00840
CISD with the most important triples (148 configurations)	-459.18771209	-0.00003

Relative importance of the triple, symmetry adapted, clusters in the valence shell of the Cℓ KL 3s3p^6 ^{2}S (C. A. Nicolaides and G. Aspromallis, unpublished). The construction of angular momentum states for the extremely complex cases of correlation vectors with many-open shells was carried out by the Bartlett-Condon-Beck (BCB) method (D. R. Beck et al., Phys. Rev. A.28, 2634, (1983)). For the ground state of Cℓ, KL 3s^{2}3p^5 ^{2}P^0, the relative contribution of the triple clusters to the correlation energy is negligible, due to the nonexistence of hole-filling correlations. [For the Fermi Sea, the MC function was 0.730 |3s3p^6⟩ + 0.683 |3s^{2}3p^{4}3d⟩ - 0.006 |3s^{2}3p^{4}4s⟩.]

Table 2. Oscillator Strengths for the Boron Photoexcitation 2s^{2}2p ^{2}P^0 → 2s^{2}ns, 2s2p^2 ^{2}S (in the length formulation).

	CI (38)	State Specific Theory		Experiment
		HF	Correlated	
3s	0.0695	0.051	0.074	0.084 (39)
4s	0.0314	0.0087	0.017	
5s	0.0038	0.0032	0.012	
6s	0.0089	0.0015	0.017	
2s2p^2	0.0004	0.091	0.014	0.035 (40)
7s	0.0033	0.0008	0.0084	

$$Z \quad : \quad \frac{1}{\sqrt{2}} \left(7a'^2 + 2a''^2 \right)$$

$$V \quad : \quad \frac{1}{\sqrt{2}} \left(7a'^2 - 2a''^2 \right)$$

In the 90° conformation (D_{2d}) these states possess no permanent dipole moment. However, pyramidalization of one of the C atoms (i.e. bending of one of the CH_2 groups) gives rise to a sudden polarization by breaking the symmetry (19,20). The suddenness and magnitude of this polarization has been investigated via large (5,000-6,000 configurations) CI with a variety of basis sets (19,20). Figure 1 shows our results (7) from a 6 x 6 state-specific, nonorthonormal CI(NONCI) and compares them with those of the previous ordinary CI calculations. The agreement is very good and suggests similar descriptions for other symmetry breaking or valence-Rydberg mixing situations in molecules. (Also shown are the results of a 6 x 6 CI using the same configurations but with an orthonormal basis set. It can be seen that they are totally wrong). Of course, larger NONCI calculations will increase the accuracy--and the magnitude of the expense in time and money. In this respect, large NONCI can certainly benefit from vectorization.

C.   Hartree-Fock Calculations of Highly Excited States

Emphasis on state-specificity implies that, in certain cases, even a Hartree-Fock function optimized for the excited state of interest can yield direct and meaningful information--thus by-passing the necessity of complex many-electron calculations. This is shown in the following three examples.

It is well known that in order to obtain one electron binding energies (BE) of inner electrons, relaxation, correlation and relativistic effects must be accounted for. This can be done with reasonable accuracy for small atoms and molecules (1,4,43-45). The solid state theory of and standard computational techniques for metals does not allow similar calculations. We (42) computed the 1s BE of Be metal in the following way: A $(Be)_{13}$ cluster was found to represent the metal well (in terms of the band structure). The expectation then was that all the local (atomic) correlation and relativistic effects could be accounted for by a many-electron atomic calculation (4), while the atomic and solid relaxation effects could be incorporated via a ΔSCF calculation. The ΔSCF calculation requires the calculation of state specific Hartree-Fock excited state of the cluster, where the central atom has a hole in the K shell. The cluster ΔSCF was found to be (42) 11.50 eV. When the local correlation and relativistic effects (4) are added (0.4 eV) the final result is 115.4 eV.

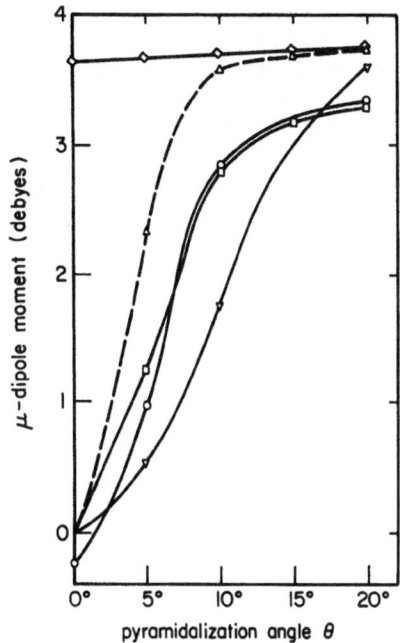

Figure 1. Increase of the dipole moment in the Z excited state of ethylene as a function of the pyramidalization angle θ as found from different CI calculations: □ Ref. 19, large CI. *Ref. 20, large CI with MO's from the $^3A''$ state. O Ref. 20, large CI with MO's from the S_1 state. ▲ Ref. 7, 6 x 6 Nonorthonormal state-specific CI. X Ref. 7, 6 x 6 standard CI using the same configurations as the 6 x 6 NON-CI. The 6 x 6 NONCI predicts a faster rise of polarization but essentially agrees with the much larger (order of 5000 x 5000) CI. The 6 x 6 standard CI yields wrong results for small angles, where it cannot restore the symmetry of the D_{2d} group. For angles around and larger than 20°, the mixing is reduced, the Z state is described well by MOs from the $7a'^2$ configuration and the dipole moment becomes stabilized in all theories.

The experimental values from the bulk are (46,47) 115.2-115.6 eV, in excellent agreement with our theory.

Recently (48,49) a model of reactivity of the $H_2B^1\Sigma_u^+$ excited state was proposed and applied for the prediction of a bound excited state ($^1A'$) of the well-studied nonreactive systems H_2+He, Ne, Ar and H_2. This model was developed after the recent MRD-CI results on HeH$_2$* (50) were interpreted. The electronic structure of these open shell $^1A'$ excited states suggested that an open-

shell, state-specific SCF calculation would be sufficient to predict binding. The SCF calculations were carried out (48) using
a modified program of W. J. Hunt which is based on the "ortho-
gonality constrained basis set expansion method" (51). Figure 2
presents a one-dimensional aspect of the energy surface of HeH_2.
A correlated, adiabatic version can be found in (50).

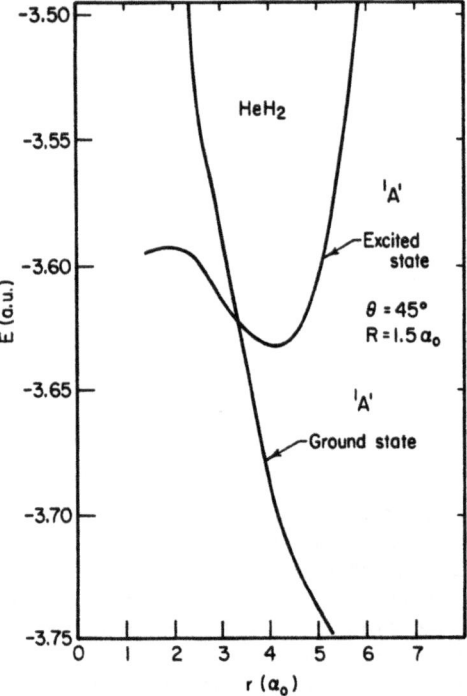

Figure 2. State-specific, open-shell SCF potential energy curves
for the first excited $^1A'$ state of HeH_2, which is found to be
bound (48), in agreement with the large CI calculations of ref. 50.

    When a single, orthonormal basis set is employed for the
description of the ground as well as excited excited states, a
single configurational description yields zero for the transition
probability to doubly or multiply excited levels. Similarly,
the algorithm of the RPA yields zero information and one must go
to "two particle-two holes" and higher order terms to obtain a
non-zero answer. These facts have caused statements in the
literature which attribute the process of multielectron excitations exclusively to electron correlation. However, when a
state-specific HF function is obtained for the multiply excited
state, relaxation of the SCF orbitals gives rise to nonorthonormality which results in a nonzero answer even at the zeroth

order, HF level. As it turns out from a variety of cases (8),
this answer, apart from its qualitative value, also has a semi-
quantitative one. Table 3 shows HF as well as correlated and
experimental results for the simultaneous excitation of the two
1s electrons. Similar calculations for the (1s2s) pair of
electrons are presented in ref. 4. I note that, although by
now there are a few atomic cases where such a HF transition
theory has been applied, the molecular domain is still unex-
plored.

D. Choice of Basis Sets and the Calculation of Total Energies

The four examples which follow are taken essentially from
different "fields" of electronic structure studies. They under-
line the importance of separating the total function space into
two parts, the zeroth order multiconfigurational Fermi-Sea and
the correlation, and of choosing and optimizing each part
separately, using HF or MCHF vector as the zeroth order solution.

Second order perturbation theory provides a simple expression for the calculation of most of the correlation energy in
not too large systems. Systematic calculations have been carried out for a number of atoms and molecules (e.g. 25, 30, 53-58).
During their current work on clusters of solid CH_4, Beck and
Kunz (30) examined electron correlation in a single CH_4 molecule
with emphasis on small, optimized atomic basis sets (1) for use
in the cluster calculations. By initializing a few primitive
virtual GTOs according to the scheme of ref. 1 ($\langle r \rangle_{GTO} = \langle r \rangle$
of carbon 2s and 2p functions) and by subsequently optimizing
them by minimizing the valence part of $E^{(2)}$, they have produced
correlation energies which surpass those of previous calculations on $E^{(2)}$ (57,58) by 1.23 eV for the valence electrons and
0.9 eV for all electrons (see Table 4). The percent increase of
$E^{(2)}$ using the small but (valence shell) optimized basis set
calculation over the standard type with large basis sets is 33%
for the valence electrons (comparison with (58)) and 17% for
all the electrons (comparison with (57)).

The smallness of Be and its valence structure are attractive
characteristics for the rigorous application and testing of the
flexibility and utility of many-body theories. In Table 5 we
present four theoretical values (4,59,60) for the binding energy
of the 1s electron and an experimental one (61). Our value
is obtained from a state-specific, 30-terms configuration-
interaction calculation on the excited state, with a $0.9409|1s2s^2\rangle +$
$0.3387|1s2p^2\rangle\ ^2S$ MCHF zeroth order vector obtained numerically (26).
As the ground state energy we used the "exact" value from Bunge's
calculation (62). A 139-terms CI of the standard type for the

Table 3. The Inclusion of Nonorthonormality in Multielectron Transition Probability Calculations.

		Hartree-Fock			Hartree-Fock Plus Correlation	
	f_L	f_V	f_A	f_L	f_V	f_A
He $\quad 1s^2 \rightarrow 2s2p\ {}^1P^0$	0.032	0.0065	0.0027	0.0084	0.0053	0.0033
Li $\quad 1s^2 2s \rightarrow 2s^2 2p\ {}^2P^0$	0.0012	0.00064	0.00022	0.00028	0.00059	0.00026

Oscillator strengths for K-shell two-electron excitations to autoionizing states (only the localized part – see Ref. 5). Length (f_L), velocity (f_V) and acceleration (f_A) form (see Ref. 2) results are given. The experimental value (52) is 0.0053.

Table 4.   Second Order Perturbation Theory Electron Correlation
            Energies of the Ground State of CH_4

	Large Basis Set(57)	Large Basis Set(58)	Small Basis Set with Valence Shell-Optimized Virtual Orbitals(30)
All Electrons	5.164 eV		6.052 eV
Valence		3.736 eV	4.969 eV

Table 5.   The 1s Binding Energy of Be.

Experiment(61)	123.6±0.1
State-specific 30-terms CI(4)	123.705
Standard, 139-terms CI(59)	123.866
"Extended Koopmans"(59)	127.922
Green's Function(60)	124.50

Energies are in eV. For the CI results, the "exact" nonrelativistic ground state energy (62) has been used. An amount of
0.027 eV should be added to the theoretical values for the relativistic contributions.

excited state (59) yields an energy higher than ours by 0.16 eV.
The proposed "extended Koopmans" method, where only the ground
state correlated function is used (59) does not do well--at least
in Be. Apparently relaxation (which is included automatically
in.a state-specific calculation) is not well-accounted for.
Finally, the last result comes from an application of a two
particle-two hole Greens' function technique (60). The deviation
from the experimental value is attributed by the authors to basis
set limitations.

    In the past two decades, a number of formalisms as well as
computational methods have been applied to the qualitative understanding of electron scattering resonances (For references, see

Refs. 3, 5 and 21). In most approaches, whether of the bound
state or scattering state point of view, emphasis is given on the
algorithms involved, without a simultaneous analysis of the
electronic structure. Thus, the basis set "problem" enters all
calculations which are based on approaches such as close-coupling,
R-matrix, complex coordinate rotation, (P,Q) projection operators
etc. A case in point is the He$^-$ $^2p^0$ resonance observed at
20.30 eV (63). Recently (64-66) there were reports of bound
state type calculations using expansions of the order of 50 terms.
In the first case (64), in the words of Bhatia and Temkin (66),
"a particularly poor choice of nonlinear parameters completely
suppressed the resonance"! Our approach to the calculation of
resonances has been guided by the ideas of state-specificity reviewed in this article. Thus, a significantly successful application was that to the He$^-$ $^2p^0$ resoance (3) where only a 2-term
MCHF vector gave a slightly better result (see Table 6) for the
position of the resonance than the previously mentioned expansions,
while a 32-term optimized wavefunction yields 20.31 eV, in agreement with experiment (see Table 6).

The traditional wisdom of quantum chemistry says that
Hylleraas basis sets converge much faster than one electron
basis sets which result in slowly converging CI expansions,
especially for "tight" pairs. A few years ago (68) correlation
optimized calculations on the $1s^2$ pair indicated that, to a
reasonable degree of accuracy, one electron functions are efficient. A similar situation was observed later on for the negative ion H$^-$ (69) and other systems (1).

Apart from the ordinary ground state, Hylleraas basis sets
have also been employed for the calculation of doubly excited
autoionizing states in He (70), in conjunction with appropriate
projection operators. Recently (67,71) we looked at states such
as He $2p^2$ ^1D, $2p3p$ $^{3,1}D$, $2s3d$ $^{3,1}D$ because of related interesting
problems of measurements of lifetimes and autoionization widths
(72). Following our approach, we first obtained the MCHF, Fermi-
Sea vectors composed of the $2p^2+2p3p+2s3d$ ^1D and $2p3d+2s3d$ ^3D
configurations. The correlation function space was optimized
variationally. As Table 6 shows, our results using 21 to 46 term
expansions are nearly the same as those obtained from a 112-term
Hylleraas expansion (70).

These findings suggest that even for certain doubly excited
states, which are supposed to present difficulties due to strong
correlation effects, the state-specific approach reviewed in this
article yields accurate energies and widths (71,73). The resulting wavefunctions can be more compact, physically more transparent
and computationally more manageable and useful than those of the
Hylleraas type which use arbitrary and large basis sets.

Table 6.   Energies (a.u.) of the First Five $^{3,1}D$ Doubly Excited
            States of He.

	State Specific Theory (67)		Standard CI with 112 Hylleraas Terms (70)
	MCHF[a]	CI	
$3D^1$	-0.5493	-0.5561(46 terms)	-0.5564
$2D^3$	-0.5540	0.5599(21 terms)	-0.5607
$2D^1$	-0.5632	-0.5684(40 terms)	-0.5693
$1D^3$	-0.5827	-0.5837(32 terms)	-0.5838
$1D^1$	-0.6970	-0.7019(46 terms)	-0.7028

[a] The Fermi-Sea, MCHF vectors are:

$$1D^1 = 0.8952|2p^2) - 0.3381|2p3p) + 0.2904|2s3d)$$
$$1D^3 = 0.8846|2p\,p) - 0.4663|2s3d)$$
$$2D^1 = 0.8813|2p3p) - 0.3812|2s3d) + .2795|2p^2)$$
$$2D^3 = 0.8695|2s3d) + 0.4939|2p3p)$$
$$3D^1 = 0.9030|2s3d) + 0.4197|2p3p) - 0.0917|2p^2)$$

Apparently, in describing the radially large doubly excited states,
the systematically chosen one-electron function spaces are more
effective than the r_{12}^n - type terms, which were introduced into
quantum chemistry to account for short range correlations.

V.   CONCLUSION

    If, in conjunction with suitable and systematic computational
methods, attention is given to the choice and optimization of
function spaces describing atoms, molecules and solids, especially
when they are in excited states, the accuracy of the description
of the various electronic structure properties increases while
the magnitude of the many-electron problem is reduced significantly.
The approach which was outlined here has been applied consistently
over the years to a variety of systems in ground and excited
states with results which agree with much larger computations
and/or with experimental values--when these are available. Its
basic elements are: a) The emphasis on the state-specific calculation of the HF or the MCHF zeroth order vector especially for
highly excited states, where gross errors in the zeroth order
description are difficult to correct efficiently via the straight-

forward incorporation of electron correlation. b) The calculation of the part beyond HF or MCHF is accomplished by incorporating the various types of localized and asymptotic electron correlation according to the zeroth order electronic configuration(s)
and the relative position of the state of interest in the energy
spectrum. c) Given the previous analysis of electronic structure and the characteristics of the observable, the aim is for
a property-specific overall calculation.

REFERENCES

1. Beck, D. R. and Nicolaides, C. A., in "Excited States in
    Quantum Chemistry", eds. Nicolaides, C. A. and Beck, D. R.,
    Reidel (1978), p. 105 and p. 329.

2. Nicolaides, C. A. and Beck, D. R., in "Excited States in
    Quantum Chemistry", eds. Nicolaides, C. A. Beck, D. R.,
    Reidel (1978), p. 143.

3. Komninos, Y., Aspromallis, G. and Nicolaides, C. A., Phys.
    Rev. A27, 1865 (1983).

4. Nicolaides, C. A., Komninos, Y. and Beck, D. R., Phys. Rev.
    A27, 3044 (1983).

5. Nicolaides, C. A. and Beck, D. R., Int. J. Quant. Chem. 14,
    457 (1978); Nicolaides, C. A., Komninos, Y. and Merkouris,
    Th., J. Quant. Chem. S15, 355 (1981).

6. Nicolaides, C. A. and Theodorakopoulos, G., Int. J. Quant.
    Chem. S14, 315 (1980).

7. Petsalakis, J., Theodorakopoulos, G., Nicolaides, C. A.,
    Buenker, R. J. and Peyerimhoff, S. D., "Non-orthonormal CI
    in Excited States. I: The Sudden Polarization Effect in
    Ethylene", to be published.

8. Nicolaides, C. A., Chem. Phys. Letts. 101, 435 (1983).

9. Weiss, A. W., Phys. Rev. 188, 119 (1969).

10. Buenker, R. J. and Peyerimhoff, S. D., in "Excited States in
    Quantum Chemistry" eds. Nicolaides, C. A., Beck, D. R.,
    Reidel (1978) pp. 45-103.

11. Kelly, H. P. and Carter, S. L., Phys. Scripta 21, 448 (1980).

12. Robb, M. A., Hegarty, D. and Prime, S., in "Excited States in Quantum Chemistry" eds. Nicolaides, C. A., Beck, D. R., Reidel (1978) p. 297.

13. Von Niessen, W., Cederbaum, L. S. and Domcke, W., in "Excited States in Quantum Chemistry" eds. Nicolaides, C. A., Beck, D. R., Reidel (1978) p. 183.

14. Linderberg, J. and Ohrn, Y., in "Excited States in Quantum Chemistry", eds. Nicolaides, C. A., Beck, D. R., Reidel (1978) p. 35 and p. 317.

15. W. Meyer, J. Chem. Phys. 64, 2901 (1976); 58, 1017 (1973).

16. Burke, P. G. and Robb, W. D., Adv. At. Mol. Phys. 11, 143 (1975).

17. Berrington, K., Burke, P. G. and Kingston, A. E., J. Phys. B8, 1459 (1975).

18. Nitzsche, L. E. and Davidson, E. R., J. Chem. Phys. 68, 3103 (1978).

19. Brooks, B. R. and Schaefer, H. F., J. Am. Chem. Soc. 101, 307 (1979).

20. Bonacic-Koutecky, V., Buenker, R. J. and Peyerimhoff, S. D., J. Am. Chem. Soc. 101, 5917 (1979).

21. Nicolaides, C. A., Phys. Rev. A6, 2078 (1972).

22. Nicolaides, C. A. and Beck, D. R., J. Phys. B6, 535 (1973).

23. Nicolaides, C. A. and Beck, D. R., Phys. Rev. A18, 1307 (1978).

24. Silverstone, H. J. and Sinanoglu, O., J. Chem. Phys. 44, 1898 (1966); Sinanoglu, O., J. Chem. Phys. 36, 706 (1962).

25. Sinanoglu, O. and Brueckner, K. A., "Three Approaches to Electron Correlation in Atoms", Yale Press (1970).

26. Froese-Fischer, C., Comp. Phys. Comm. 4, 107 (1972).

27. Beck, D. R. and Nicolaides, C. A., Int. J. Quant. Chem. S8, 17 (1974).

28. Nicolaides, C. A. and Beck, D. R., Chem. Phys. Lett. 36, 79 (1975).

29. Aspromallis, G., Nicolaides, C. A. and Beck, D. R., Phys. Rev. A28, 1879 (1983).

30. a) Beck, D. R., Int. J. Quant. Chem. S15, 521 (1981);
     b) Beck, D. R. and Kunz, A. B., unpublished.

31. Nicolaides, C. A., Komninos, Y. and Beck, D. R. Phys. Rev. A24, 1103 (1981).

32. Beck, D. R., Nicolaides, C. A., and Aspromallis, G., Phys. Rev. A24, 3252 (1981).

33. King, H. F., Stanton, R. E., Kim, H., Wyatt, R. E. and Parr, R. G., J. Chem. Phys. 47, 1936 (1967).

34. Schneider, B. I. and Collins, L. A., J. Phys. B15, L335 (1982).

35. Schneider, B. I., LeDourneuf, M. and Lan, V. K., Phys. Rev. Lett. 43, 1926 (1979).

36. Nicolaides, C. A. and Adamides, E., Phys. Rev. A27, 1691 (1983).

37. Beck, D. R. and Nicolaides, C. A., Phys. Lett. 61A, 227 (1977).

38. Nesbet, R. K. Phys. Rev. A14, 1065 (1976).

39. Kernahan, J. A., Pinnington, E. H., Livingston, A. E. and Irwin, D. J. G. Phys. Scripta 12, 319 (1975).

40. Martinson, I., Bickel, W. S. and Olme, A., J. Opt. Soc. Am. 60, 1213 (1970).

41. Mulliken, R. S., Phys. Rev. 43, 279 (1933).

42. Nicolaides, C. A., Zdetsis, A. and Andriotis, A., unpublished.

43. Butscher, W., Buenker, R. J. and Peyerimhoff, S. D., Chem. Phys. Letts. 52, 449 (1977).

44. Cederbaum, L. S. and Domcke, W., J. Chem. Phys. 66, 5084 (1977).

45. Beck, D. R. and Nicolaides, C. A., Phys. Rev. A26, 857 (1982).

46. Shirley, D. A., Martin, R. L., Kowalczyk, S. P., McFeely, F. R. and Ley, L., Phys. Rev. 15, 544 (1977).

47. Hochst, H., Steiner, P. and Hufner, S., Phys. Lett. 60A, 69 (1977).

48. Nicolaides, C. A. and Zdetsis, A., "Theory of Chemical Reactions of Vibronically Excited $H_2(B^1\Sigma_u^+)$: Bound Excited States of Noble Gas Dihydrides", J. Chem. Phys., in press.

49. Nicolaides, C. A., Theodorakopoulos, G., and Petsalakis, I., "Prediction of a Strongly Bound Excited State of H_4", J. Chem. Phys., in press.

50. Farantos, S., Theodorakopoulos, G. and Nicolaides, C. A., Chem. Phys. Lett. 100, 263 (1983).

51. Hunt, W. J., Dunning, T. H. and Goddard, W. A., Chem. Phys. Lett. 3, 606 (1969).

52. Madden, R. P. and Codling, K., Astr. J. 141, 364 (1965).

53. Schulman, J. M. and Kaufman, D. N., J. Chem. Phys. 53, 477 (1970).

54. Pople, J. A., Binkley, J. S. and Seeger, R., Int. J. Quant. Chem. S10, 1 (1976).

55. Krishnan, R. and Pople, J. A., Int. J. Quant. Chem. 14, 91 (1978).

56. Eggarter, E. and Eggarter, T. P., J. Phys. B11, 1157 (1978).

57. Bartlett, R. J. and Purvis, G. D., Phys. Scripta, 21, 255 (1980).

58. Frisch, M. J., Drishnan, R. and Pople, J. A., Chem. Phys. Lett. 75, 66 (1980).

59. Morrison, R. C., Chem. Phys. Lett. 62, 131 (1979).

60. Walter, O. and Schirmer, J., J. Phys. B14, 3805 (1981).

61. Bisgaard, P., Bruch, R., Dahl, P., Fastrup, B. and Robro, M., Phys. Scripta 17, 49 (1978).

62. Bunge, C. F., Phys. Rev. A14, 1965 (1976).

63. Brunt, J. N. H., King, G. C., Read, F. H., J. Phys. B10, 433 (1977).

64. Wakid, S., Bhatia, A. K. and Temkin, A., Phys. Rev. A21, 496 (1980).

65.   Chung, K. T., Phys. Rev. A23, 1079 (1981).

66.   Bhatia, A. K. and Temkin, A., Phys. Rev. A23, 3361 (1981).

67.   Aspromallis, G., Komninos, Y. and Nicolaides, C. A., J. Phys. B, in press.

68.   Beck, D. R. and Nicolaides, C. A., Int. J. Quant. Chem. S10, 119 (1976).

69.   Beck, D. R. and Nicolaides, C. A., Chem. Phys. Lett. 59, 525 (1978).

70.   Bhatia, A. K., Phys. Rev. A6, 120 (1972).

71.   Komninos, Y. and Nicolaides, C. A., Phys. Scripta 28, 472 (1983).

72.   Cederquist, H., Kisielinski, M. and Mannervik, S., J. Phys. B16, L479 (1983).

73.   Aspromallis, G. and Nicolaides, C. A., J. Phys. B16, L251 (1983).

THE TREATMENT OF ELECTRON CORRELATION: WHERE DO WE GO FROM HERE?

Isaiah Shavitt

Department of Chemistry, Ohio State University,
Columbus, Ohio 43210, U.S.A.

While substantial progress has been made in recent years in the methods and implementation of quantum chemical calculations, the results of such calculations still fall short, in many cases, of requirements. Generally, single-reference methods have difficulties in handling nondynamical correlation effects (which are due to near degeneracies of the reference function). The configuration interaction method easily deals with these effects by the use of multireference expansions, but has much greater difficulty than MBPT and coupled cluster methods in handling dynamical correlation because of its inability to treat disconnected cluster effects adequately. The most effective approach should combine a multiconfiguration reference function for the treatment of nondynamical correlation with a coupled cluster treatment (or a good approximation thereof) of dynamical correlation.

INTRODUCTION

The last ten years have seen remarkable developments in the techniques and capabilities of molecular electronic structure calculations. These developments have covered practically all aspects of such calculations, including the basic theoretical formalisms, their organization into effective computational procedures, the adaptation of such procedures to modern vector-oriented computers, and last but not least, the continuing rapid increase in the power of computers. Yet, the goal of being able to obtain accurate and reliable results for many quantites of physical and chemical interest, even for moderate size systems, has remained largely elusive.

C. E. Dykstra (ed.),
Advanced Theories and Computational Approaches to the Electronic Structure of Molecules, 185–196.
© *1984 by D. Reidel Publishing Company.*

It cannot be denied that there have been substantial improvements in the results of quantum chemical calculations. The theoretical predictions of some quantities, such as molecular geometries and energy differences for some types of molecular rearrangements, have improved to the point that they can often be considered quite satisfactory. But much too often the results still fall well short of requirements, and some molecular properties, notably electron affinities and many dissociation and excitation energies, have proved to be rather intractable. For some examples of both successes and difficulties of recent calculations we cite the extensive studies of Davidson and co-workers on the electron affinity of methylene [1] and on the electronic properties of the water molecules [2], the detailed examination of the dipole and quadrupole moments and polarizabilities of Be, HF, and H_2O by Diercksen, Roos, and Sadlej [3], and the systematic studies of dissociation energies by Pople et al. [4] and by Binkley and Frisch [5].

In recent years we have seen basis sets expanding in size and type, including f and even g functions in some cases, the lengths of CI expansions increasing to hundreds of thousands, or even more than a million, terms, MBPT treatments expanding to incorporate all fourth-order terms, and coupled cluster methods being extended to include additional types of clusters. Nevertheless, we usually find that we need still larger basis sets and more complete treatments of electron correlation, with the concomitant massive increase in computational power that this entails, if we are to obtain entirely satisfactory results. Short of a breakthrough with an entirely new approach (perhaps one which does not require basis sets, such as the quantum Monte Carlo method [6]), there still is much to be done in continuing the development of the traditional methods and improving the computational capabilities.

While basis set problems are emerging as the most severe limitations on the accuracy of quantum chemical calculations in many cases, we shall focus here on the treatment of electron correlation within the given basis, and particularly on the advantages and limitations of the configuration interaction approach.

DYNAMICAL AND NONDYNAMICAL ELECTRON CORRELATION

The electron correlation effects in atomic and molecular systems are usually defined [7] in terms of the difference between the Hartree-Fock (HF) description of the electronic state and the exact solution of the nonrelativistic Schrödinger equation. It is useful to distinguish two types of electron correlation, dynamical and nondynamical [8-10], even though this distinction is not always clearcut.

Dynamical correlation represents the intuitive idea of the Coulomb-induced correlation between the motions of the different electrons. Nondynamical correlation reflects other deficiencies of the Hartree-Fock description (other than the effective averaging of the interelectron repulsions), which make it necessary to include more than one configuration in a satisfactory zero-order description of the wave function. This is generally due to near degeneracies, such as that involving the $1s^2 2s^2$ and $1s^2 2p^2$ configurations in the beryllium atom, and includes the possible rearrangement of electrons in partly filled subshells and the improper Hartree-Fock description of molecular dissociation in the treatment of potential energy surfaces. (The use of unrestricted Hartree-Fock, or UHF, wave functions can alleviate these problems in some cases, but does not provide a general solution and has other disadvantages.)

The most generally effective approach for dealing with nondynamical correlation is the use of a multiconfigurational zero-order function, particularly a multiconfigurational Hartree-Fock (MCHF or MCSCF) wave function. But it would be a mistake to expect this approach, by itself, to provide a satisfactory treatment of the full effects of dynamical correlation. To provide such a treatment would normally require the cumulative effect of very large numbers of configurations, in order to make up for the lack of explicit interelectron coordinates in the wave function.

Three methods have emerged preeminent for the treatment of dynamical correlation: configuration interaction (CI), many-body perturbation theory (MBPT, also referred to as Møller-Plesset theory), and the coupled cluster approach (CCA). Approximate versions of these, such as contracted CI [11-15] and the various CEPA (coupled electron pair approximation) and other versions of CCA [11,16-21], have also been used. Each of these approaches has its strengths and weaknesses, relating either to problems of implementation and generality or to intrinsic limitations of implementable procedures. The principal problem of MBPT and CCA is the difficulty of implementing effective multireference versions, such as quasidegenerate MBPT [22-28] or multiconfiguration based forms of CCA [21,25,29-34]. The most serious deficiency of CI, contributing to its agonizingly slow convergence, is the lack of extensivity (proper dependence of the energy on the size of the system [35-40], closely related to size consistency [41]) of truncated CI expansions. This is discussed more fully in the next section.

CONFIGURATION INTERACTION

Modern configuration interaction techniques are generally based on the direct CI approach of Roos and Siegbahn [42,43]. This approach bypasses the explicit construction of the Hamiltonian matrix and (like many-body methods) works directly with an ordered list of the molecular orbital integrals, incorporating the determination of matrix element contributions of these integrals directly into the iterative procedure for the solution of the matrix eigenvalue problem. In this way it eliminates serious data storage and handling problems, and significantly reduces the amount of computational overhead in the calculation.

The original form of the direct CI approach [42] was limited to closed-shell single-reference cases, but new techniques which exploit the systematics of the Hamiltonian matrix structure in more general cases have now extended the capabilities of direct CI to general multireference CI (MR-CI) expansions. Three basic forms (and some variations) of these direct MR-CI techniques have emerged, though they share some aspects in their implementation. One of these is the symbolic matrix approach of Liu and Yoshimine [44], in which prototype matrix element formulas are generated, based on the pattern of orbital occupancies in representative configuration state functions. An important strength of this approach is its ability to handle symmetry adaptation for non-Abelian point groups. (A pattern-based approach has also been implemented in a conventional CI context by Buenker [45]). A second form is the matrix-formulated direct CI procedure [14,19,40,46-49], based on the SCEP (self-consistent electron pairs) formalism [46-48]. The third form is based on the graphical unitary group approach (GUGA) [50-63]. These and related approaches have greatly expanded the capabilities of CI treatments and have made possible the use of CI expansions containing more than a million configuration state functions [60,64] (for an analysis of modern direct CI algorithms see Saunders and van Lenthe [65]).

The most important advantage of the CI approach is its flexibility. In the context of conventional CI [66], in which the Hamiltonian matrix is constructed explicitly in an ordered form (e.g., by rows of the lower triangle), it is generally possible to accommodate any choice of configuration state functions (CSF), based on arbitrary sets of reference configurations [66-70] and any desired excitation levels, or on any other criteria. It also allows individual selection of CSF based, for example, on energy contributions [66-72], and extrapolation of results at several selection thresholds to zero threshold [70,72-74]. This flexibility allows specific adaptation of the CI expansion to various requirements, including the simultaneous calculation of several electronic states [66,69,70,75].

Unfortunately, some of this flexibility is given up in the direct CI approach, due to the need to exploit regularities in the structure of the Hamiltonian matrix which are based, in turn, on regularities in the list of CSF. The choice of reference configurations is still essentially arbitrary (though choices which involve a large number of variably occupied orbitals can lead to excessively long lists of prototype formulas and to other difficulties). However, the maximum excitation level relative to the reference configurations is usually limited to double excitations for viable treatments (but see Saxe et al. [60]). Individual configuration selection is generally impractical, though selections of sets of configurations which share the same "internal" orbital structure and range over all compatible "external" orbital occupancies is viable and has been implemented [56,76]. Such selection capabilities are very important, since they can reduce extremely long CI expansions to managable size without excessive impact on the results.

The ability to choose arbitrary sets of reference configurations is the key for the effectiveness of MR-CI techniques in handling nondynamical correlation. Because of this they provide the means for the balanced treatment of different regions of a potential energy surface, including the proper description of molecular dissociation. In fact, with suitable choices of reference configurations it is possible to obtain computed potential energy surfaces in which the error, relative to the corresponding full CI surface, is almost constant [77]. However, practical considerations greatly limit the number of reference configurations which can be employed.

It has been our experience [76,77] that the number of CSF generated by different sets of reference configurations in a given basis set is almost proportional to the number of reference configurations. This often rules out the use of complete active space (CAS) reference sets in MR-CI calculations, but this problem can be overcome, at least to some extent, by the use of "internally contracted" CI expansions [11,13,14] and by methods which optimize the external (correlating) orbitals [49], based on the nonorthogonal pair-natural-orbitals (PNO) approach of Meyer [11,17]. Alternatively, CSF selection by objective criteria (CI coefficient size or energy contribution) can be used both at the reference configuration level and the MR-CI level to reduce the size of the expansion. An example of this approach from work by Brown and Shavitt [76] on the oxygen molecule is shown in Table 1.

Table 1.   MCSCF and MR-CI calculations on O_2 in a (5s4p2d) STO basis (R = 2.2819 bohr), using D_{2h} symmetry (from Ref. [76]).

Reference space[a]	No. of CSF		Energy (hartrees)	
	Ref.	CI	MCSCF	CI
Hartree-Fock	1	12007	−149.657831	−150.050061
CAS $(3\sigma_g,3\sigma_u,1\pi_g,1\pi_u)$	14	164296	−149.757764	−150.076166
CAS (as above + $2\pi_u$)	298	~4000000[b]	−149.804803	
Ref. selection (0.5 mh) 43	43	767514	−149.800104	
CI selection (0.5 μh)		253572		−150.079118
Ref. selection (0.25 mh) 72	72	1108204		
CI selection (0.5 μh)		295623		−150.0793

[a]The complete active space reference space is described in terms of the "active" (variably occupied) orbitals.   The $1\sigma_g$, $1\sigma_u$, $2\sigma_g$,,and $2\sigma_u$ orbitals are doubly occupied in all reference configurations.   The $1\sigma_g$ and $1\sigma_u$ orbitals are doubly occupied in all CSF in the CI expansion.   The selection thresholds for energy contributions (relative to the 298 reference configuration case) are given for both reference configuration selection (in millihartrees) and CI selection (in microhartrees).   The CI selection threshold is for the aggregate energy contribution of the set of all CSF which share the same internal structure.

[b]Estimated

     As noted previously, the principal deficiency of the CI approach is the lack of extensivity or size consistency of truncated CI expansions.   Not only does this contribute to the very slow convergence of the CI method, but it is also respon-sible, at least in part, for the great difficulty in computing dissociation energies and related quantities by CI techniques. While the use of multireference expansions can alleviate this problem substantially in some cases [77], differences in the dynamical component of the correlation energy between a molecule and its fragments would tend to make it extremely difficult to obtain accurate dissociation energies without the use of imprac-tically large reference configuration sets.   For example, in the O_2 study discussed in Table 1 [76], the dissociation energy computed in the selected MR-CI calculation with 43 reference CSF (the selections were carried out at several representative internuclear distances, and the selected CSF sets were merged) was 0.43 eV short of the experimental value. While the major part of this error is probably due to basis set deficiencies (including the lack of f functions), some of it is undoubtedly due to the limitations of the CI expansion and the concomitant deficiency in the description of dynamical correlation within the given basis. Furthermore, expanding the basis set would make adequate CI treatments even less feasible. This problem is likely to be aggravated very substantially when dealing with bigger molecules [78].

The origin of the difficulty which CI has in dealing with dynamical correlation is in the inability of truncated CI to account for the correlation effects of "disconnected clusters" [21,36-40]. While electrons correlate primarily in pairs, accounting for the central role of double excitations in CI, the simultaneous, but independent, correlation of different electron pairs (disconnected pairs) is also important, and increases rapidly in importance as the number of electrons in the system increases. It is the omission of such correlations, which cannot be separated in a CI treatment from connected higher-order clusters, and can only be accounted for by including higher excitation configurations in the expansion, which is responsible for the lack of extensivity of truncated CI, and therefore for its slow convergence in treating dynamical correlation. The MBPT and, particularly, the CCA methods can take advantage of the relatively simple structure of the disconnected cluster terms in order to include their effects without including the corresponding connected terms and without a prohibitive increase in the size of the calculation, and that is why they can be much more effective than CI in dealing with dynamical correlation.

Several approximate schemes exist for correcting truncated CI treatments for the effects of disconnected clusters. These include the relatively simple "Davidson correction" [78] and its many variations (for a recent discussion see Paldus [21]), as well as the more complicated CEPA and related methods [11, 16-21,79-82], which require recalculation of modified CI vectors and are mostly related to the coupled cluster approach. The Davidson correction, which is essentially a rescaling of the double excitation contribution in a truncated CI expansion, was developed for single-reference expansions. It has also been used for multireference cases (e.g., [83]) with some success, but without theoretical justification. The CEPA methods were originally derived for single-reference expansions, though multireference versions have been proposed [80-82] and are likely to prove very useful. A well founded, proven reliable, and relatively simple disconnected cluster correction scheme for multireference CI calculations should be a very important contribution at this point.

WHERE DO WE GO FROM HERE?

While MR-CI techniques can be very effective in dealing with nondynamical correlation, their neglect of disconnected cluster contributions makes their treatment of dynamical correlation unsatisfactory, and leads to extensivity and size consistency problems. Ad hoc corrections for disconnected cluster effects can alleviate these problems (at least for single-reference expansion), but can only be considered as a partial solution. Without the use of impractically large sets of reference configurations, we cannot expect the multireference nature of the CI expansion to compensate for the inadequacies in the treatment of dynamical correlation in CI.

On the other hand, traditional MBPT and CCA procedures are limited to single-reference treatments, and thus have great difficulty in dealing with nondynamical correlation effects (in this respect CCA can be superior to finite-order MBPT in handling near-degeneracies of the reference function [21,38,84-86], but has the disadvantage that it is more difficult to include triple excitation terms in CCA [87]). Extensions of these approaches to multiconfigurational reference functions [21-34] have proved difficult to formulate and even more difficult to implement and apply. Nevertheless, such extensions must be considered the most promising avenue for further progress towards reliable ab initio calculations of molecular electronic wave functions and properties.

It should be noted that if a quasidegenerate (or other multireference) form of MBPT is to be used, it is unlikely that less than a fourth-order treatment will be satisfactory, just as in the case of ordinary MBPT. This is because, as in MR-CI, any reference space of practical size cannot be expected to contribute much, by itself, to dynamical correlation; its primary purpose is to deal with nondynamical effects. Flexibility in the choice of the reference space is also important, in order to be able to keep its size as small as possible and to reduce convergence problems due to "intruder states" [25].

Ultimately, a multireference CCA formalism (which represents an infinite-order summation of certain MBPT terms) is probably the single most promising approach. The first steps in this direction [29-34] show that the difficulties are likely to be overcome, and it is to be hoped that the approximations which it has been found necessary to make in the early treatments will gradually be eliminated. The result would be a method which combines the effectiveness of multireference approaches in treating nondynamical correlation with the great efficiency of CCA in handling dynamical correlation effects.

While such a development may eventually eclipse CI, the development of effective cluster correction schemes for MR-CI expansions may greatly extend the usefulness of the conceptually simple and attractive CI approach.

ACKNOWLEDGMENT

This work was supported by the National Science Foundation under Grant CHE-8219408.

REFERENCES

1. Feller, D., McMurchie, L.E., Borden, W.T., and Davidson, E.R.: 1982, J. Chem. Phys. 77, p. 6134.
2. Davidson, E.R. and Feller, D.: 1984, Chem. Phys. Letters 104, p. 54.
3. Diercksen, G.H.F., Roos, B.O., and Sadlej, A.J.: 1983, Int. J. Quantum Chem. Symp. 17, p. 265.
4. Pople, J.A., Frisch, M.J., Luke, B.T., and Binkley, J.S.: 1983, Int. J. Quantum Chem. Symp. 17, p. 307.
5. Binkley, J.S. and Frisch, M.J.: 1983, Int. J. Quantum Chem. Symp. 17, p. 331.
6. Reynolds, P.J., Ceperley, D.M., Alder, B.J., and Lester Jr., W.A.: 1982, J. Chem. Phys. 77, p. 5593.
7. Löwdin, P.-O.: 1959, Adv. Chem. Phys. 2, p. 207.
8. Sinanoğlu, O.: 1961, Proc. Nat. Acad. Sci. USA 47, p. 1217.
9. Silverstone, H.J. and Sinanoğlu, O.: 1966, J. Chem. Phys. 44, p. 1899.
10. Hollister, C. and Sinanoğlu, O.: 1966, J. Am. Chem. Soc. 88, p. 13.
11. Meyer, W.: 1977, in "Methods of Electronic Structure Theory," edited by H.F. Schaefer III (Plenum, New York), p. 413.
12. Siegbahn, P.E.M.: 1977, Chem. Phys. 25, p. 197.
13. Siegbahn, P.E.M.: 1980, Int. J. Quantum Chem. 18, p. 1229.
14. Werner, H.-J. and Reinsch, E.-A.: 1982, J. Chem. Phys. 76, p. 3144.
15. Siegbahn, P.E.M.: 1983, Int. J. Quantum Chem. 23, p. 1869.
16. Meyer, W.: 1971, Int. J. Quantum Chem. Symp. 5, p. 341.
17. Meyer, W.: 1973, J. Chem. Phys. 58, p. 1017.
18. Ahlrichs, R., Lischka, H., Staemmler, V., and Kutzelnigg, W.: 1975, J. Chem. Phys. 62, p. 1225.
19. Ahlrichs, R.: 1979, Comput. Phys. Commun. 17, p. 31.

20. Cullen, J.M. and Zerner, M.C.: 1982, J. Chem. Phys. 77, p. 4088.
21. Paldus, J.: 1983, in "New Horizons of Quantum Chemistry," edited by P.-O. Löwdin and B. Pullman (Reidel, Dordrecht, Holland), p. 31.
22. Brandow, B.H.: 1967, Rev. Mod. Phys. 39, p. 771.
23. Brandow, B.H.: 1977, Adv. Quantum Chem. 10, p. 187.
24. Brandow, B.H.: 1979, Int. J. Quantum Chem. 15, p. 207.
25. Brandow, B.: 1983, in "New Horizons of Quantum Chemistry," edited by P.-O. Löwdin and B. Pullman (Reidel, Dordrecht, Holland), p. 15.
26. Lindgren, I.: 1974, J. Phys. B 7, p. 2441.
27. Kvasnička, V.: 1977, Adv. Chem. Phys. 36, p. 345.
28. Hose, G. and Kaldor, U.: 1979, J. Phys. B 12, p. 3827.
29. Mukherjee, D., Moitra, R.K., and Mukhopadhyay, A.: 1975, Mol. Phys. 30, p. 1861.
30. Lindgren, I.: 1978, Int. J. Quantum Chem. Symp. 12, p. 33.
31. Jeziorski, B. and Monkhorst, H.J.: 1981, Phys. Rev. A 24, p. 1668.
32. Banerjee, A. and Simons, J.: 1981, Int. J. Quantum Chem. 19, p. 207.
33. Kvasnička, V., Laurinc, V., Biskupič, S., and Haring, M.: 1983, Adv. Chem. Phys. 52, p. 181.
34. Laidig, W.D. and Bartlett, R.J.: 1984, Chem. Phys. Letters 104, p. 424.
35. Primas, H.: 1965, in "Modern Quantum Chemistry," edited by O. Sinanoğlu (Academic, New York), Vol. 2, p. 45.
36. Meunier, A., Levy, B., and Berthier, G.: 1976, Int. J. Quantum Chem. 10, p. 1061.
37. Sasaki, F.: 1977, Int. J. Quantum Chem. Symp. 11, p. 125.
38. Bartlett, R.J. and Purvis, G.D.: 1978, Int. J. Quantum Chem. 14, p. 561.
39. Bartlett, R.J.: 1981, Ann. Rev. Phys. Chem. 32, p. 359.
40. Ahlrichs, R.: 1983, in "Methods in Computational Molecular Physics," edited by G.H.F. Diercksen and S. Wilson (Reidel, Dordrecht, Holland), p. 209.
41. Pople, J.A., Binkley, J.S., and Seeger, R.: 1976, Int. J. Quantum Chem. Symp. 10, p. 1.
42. Roos, B.: 1972, Chem. Phys. Letters 15, p. 153.
43. Roos, B.O. and Siegbahn, P.E.M.: 1977, in "Methods of Electronic Structure Theory," edited by H.F. Schaefer III (Plenum, New York), p. 277.
44. Liu, B. and Yoshimine, M.: 1981, J. Chem. Phys. 74, p. 612.
45. Buenker, R.J.: 1980, in "Molecular Physics and Quantum Chemistry into the '80's," edited by P.G. Burton (Department of Chemistry, University of Wollongong, Wollongong, N.S.W. 2500, Australia), p. 1.5.1.
46. Meyer, W.: 1976, J. Chem. Phys. 64, p. 2901.
47. Dykstra, C.E., Schaefer III, H.F., and Meyer, W.: 1976, J. Chem. Phys. 65, p. 2740.

48. Chiles, R.A. and Dykstra, C.E.: 1981, J. Chem. Phys. 74, p. 4544.
49. Taylor, P.R.: 1981, J. Chem. Phys. 74, p. 1256.
50. Paldus, J.: 1976, in "Theoretical Chemistry: Advances and Perspectives," edited by H. Eyring and D.G. Henderson (Academic, New York), Vol. 2, p. 131.
51. Paldus, J., and Boyle, M.J.: 1980, Phys. Scripta 21, p. 295.
52. Paldus, J.: 1981, in "The Unitary Group for the Evaluation of Electronic Energy Matrix Elements," edited by J. Hinze (Springer, Berlin), p. 1.
53. Shavitt, I.: 1977, Int. J. Quantum Chem. Symp. 11, p. 131.
54. Shavitt, I.: 1978, Int. J. Quantum Chem. Symp. 12, p. 5.
55. Shavitt, I.: 1981, in "The Unitary Group for the Evaluation of Electronic Energy Matrix Elements," edited by J. Hinze (Springer, Berlin), p. 51.
56. Lischka, H., Shepard, R., Brown, F.B., and Shavitt, I.: 1981, Int. J. Quantum Chem. Symp. 15, p. 91.
57. Shavitt, I.: 1983, in "New Horizons of Quantum Chemistry," edited by P.-O. Löwdin and B. Pullman (Reidel, Dordrecht, Holland), p. 279.
58. Brooks, B.R. and Schaefer III, H.F.: 1979, J. Chem. Phys. 70, p. 5092.
59. Brooks, B.R., Laidig, W.D., Saxe, P., Handy, N.C., and Schaefer III, H.F.: 1980, Phys. Scripta 21, p. 312.
60. Saxe, P., Fox, D.J., Schaefer III, H.F., and Handy, N.C.: 1982, J. Chem. Phys. 77, p. 5584.
61. Siegbahn, P.E.M.: 1979, J. Chem. Phys. 70, p. 5391.
62. Siegbahn, P.E.M.: 1980, J. Chem. Phys. 72, p. 1647.
63. Siegbahn, P.E.M.: 1981, in "The Unitary Group for the Evaluation of Electronic Energy Matrix Elements," edited by J. Hinze (Springer, Berlin), p. 119.
64. Liu, B.: 1983, private communication.
65. Saunders, V.R. and van Lenthe, J.H.: 1983, Mol. Phys. 48, p. 923.
66. Shavitt, I.: 1977, in "Methods of Electronic Structure Theory," edited by H.F. Schaefer III (Plenum, New York), p. 189.
67. Whitten, J.L. and Hackmeyer, M.: 1969, J. Chem. Phys. 51, p. 5584.
68. McLean, A.D. and Liu, B.: 1973, J. Chem. Phys. 58, p. 1066.
69. Kahn, L.R., Hay, P.J., and Shavitt, I.: 1974, J. Chem. Phys. 61, p. 3530.
70. Buenker, R.J., Peyerimhoff, S.D., and Butscher, W.: 1978, Mol. Phys. 35, p. 771.
71. Bender, C.F., and Davidson, E.R.: 1969, Phys. Rev. 183, p. 23.
72. Buenker, R.J. and Peyerimhoff, S.D.: 1974, Theor. Chim. Acta 35, p. 33.

73. Buenker, R.J. and Peyerimhoff, S.D.: 1975, Theor. Chim. Acta 39, p. 217.
74. Jackels, C.F. and Shavitt, I.: 1981, Theor. Chim. Acta 58, p. 81.
75. Raffenetti, R.C., Hsu, K., and Shavitt, I.: 1977, Theor. Chim. Acta 45, p. 33.
76. Brown, F.B.: 1982, Ph.D. Dissertation (Department of Chemistry, Ohio State University, Columbus, Ohio).
77. Brown, F.B., Shavitt, I., and Shepard, R.: 1984, Chem. Phys. Letters (in press).
78. Davidson, E.R.: 1974, in "The World of Quantum Chemistry," edited by R. Daudel and B. Pullman (Reidel, Dordrecht, Holland), p. 17.
79. Paldus, J., Wormer, P.E.S., Visser, F., and van der Avoird, A.: 1982, J. Chem. Phys. 76, p. 2458.
80. Wenzel, K.B.: 1982, J. Phys. B 15, p. 835.
81. Prime, S., Rees, C., and Robb, M.A.: 1981, Mol. Phys. 44, p. 173.
82. Brändas, E.J., Bendazzoli, G.L., and Ortolani, F.: 1983, Int. J. Quantum Chem. Symp. 17, p. 321.
83. Buenker, R.J. and Peyerimhoff, S.D.: 1983, in "New Horizons of Quantum Chemistry," edited by P.-O. Löwdin and B. Pullman (Reidel, Dordrecht, Holland), p. 183.
84. Jankowski, K. and Paldus, J.: 1980, Int. J. Quantum Chem. 18, p. 1243.
85. Purvis III, G.D., Shepard, R., Brown, F.B., and Bartlett, R.J.: 1983, Int. J. Quantum Chem. 23, p. 835.
86. Wilson, S., Jankowski, K., and Paldus, J.: 1983, Int. J. Quantum Chem. 23, p. 1781.
87. Lee, Y.S. and Bartlett, R.J.: 1984, J. Chem. Phys. 80 (in press).

COMPUTER TECHNOLOGY IN QUANTUM CHEMISTRY

Clifford E. Dykstra

School of Chemical Sciences
University of Illinois
505 South Mathews Avenue
Urbana, Illinois 61801

Henry F. Schaefer III

Department of Chemistry
University of California
Berkeley, California 94720

The link between computer technology and quantum chemistry
is an important one. The current fast pace of methodological
developments anticipates more powerful computing systems and
opens up more and more chemical problems to theoretical/computa-
tional investigation.

DISCUSSION

It is in conjunction with the remarkable progress in computer
technology over the last two decades that chemical theorists are
now making the sort of contributions to science envisioned by
Dirac more than fifty years ago. Looking ahead, we may be confi-
dent that by the year 2000 theoretical/computational chemistry
will be an equal partner with the traditional fields of organic,
inorganic, and physical chemistry. Clearly, computers are and
will continue to be part of the arsenal of instrumentation available to molecular science.

The field of electronic structure has always tended to challenge the capabilities of existing computer technology and has
pressed the use of computers in chemistry to its most sophisticated
level. McLean's molecular orbital calculation on the electronic

C. E. Dykstra (ed.),
Advanced Theories and Computational Approaches to the Electronic Structure of Molecules, 197–202.
© *1984 by D. Reidel Publishing Company.*

structure of acetylene (1), done without a high level computer
language and with limited memory by today's standards, stands out
as an important early example. It showed how far computers
could be pushed into giving chemists the detailed quantum mechanical understanding that they sought.

After years of reliance on large, central computer systems,
Miller and Schaefer tried a remarkable experiment in 1973. Ac-
knowledging the technical advances on the small end of the computer spectrum, they made the first attempt to use a minicomputer
to solve scientific problems that had previously been thought to
require the largest existing mainframes, e.g., the CDC 7600. The
Berkeley experiment has been documented in the open literature (2-5) and in a longer special report made available by the
National Science Foundation (6). That the minicomputer was successful for the planned calculations as well as cost-effective
was due in part to the fact that it was a dedicated computer,
meaning it was a system where resources could be arranged in the
optimal way for a small group of users with common needs.

What might seem a reversal of the trend toward dedicated
minicomputers has taken place over the last few years and is due
largely to the success of the Cray supercomputers, where we use
supercomputer to refer to a vectorized or a multiple processor
mainframe. For the past four years, a Cray system has been
available to qualified British theoretical chemists at the SRC
Daresbury Laboratory where Saunders and Guest have demonstrated
that vectorization can result in speed enhancements approaching a factor of 30 relative to non-vectorized operations on a
Cray-1. West German universities and research laboratories
already have half a dozen supercomputers in place; theoretical
chemists in Sweden will soon have access to an industry owned
Cray-1; and the Institute for Molecular Science in Okazaki,
Japan, is likely to receive one of the first of the new
Hitachi supercomputers. In the United States, supercomputers
are being used for quantum chemical calculations in national
laboratories at Los Alamos and Livermore and at the handful of
universities that have acquired these systems. This represents
a promising development for electronic structure since the ideal
computing situation for a theoretical/computational chemist during
the last half of the 1980's would be to have convenient access to
a supercomputer or its equivalent with a healthy allocation of
time, e.g., about 10 hrs./week in Cray-XMP equivalent hours.

Minicomputer system developments have certainly kept pace
with the big mainframes and we can now identify a number of systems deserving of the designation super-minicomputer. One of
these, the Floating Point Systems Model 164, is especially interesting because it can operate not only as an attached array
processor to accelerate essentially vector processes, but also,

with attached disks, it can operate with little use of the host
system. The groups of Kunz and Flynn in the Materials Research
Lab at the University of Illinois and the Dunning group at
Argonne National Laboratory have been carrying out electronic
structure calculations on the FPS systems for over a year. Both
groups have reported this system to be quite effective (7,8).
(The Argonne group's findings are presented elsewhere in this
volume.) The parallel-pipeline architecture used by FPS seems to
be consistently capable of operating at six times the speed of a
VAX 11/780, a conventional and widely used minicomputer produced
by Digital Equipment, and that enhancement is obtained just by put-
ting regular FORTRAN code through the FPS cross-compiler. Much
greater enhancements, up to 25 times the speed of the VAX, are
possible with some program restructuring somewhat along the lines
of that required for optimal use of supercomputers. Minicomputers
of conventional architecture are becoming much faster, too. For
instance, the Harris 1000 and the Gould 32/87, equipped with a
multiply instruction accelerator, are roughly five times the speed
of a VAX 11/780; and of course Digital's VAX line continues to
expand with more powerful systems. Most of the super-minicomputer
systems that are available are in the price range of several $100K,
but they all offer much more computing per dollar than the minicomputers of ten or even five years ago.

    While quantum chemists have had to pay attention to computer
technology for only the last two decades, many of the ideas of
electronic structure go back further, predating the capability
for implementation and real application. For more than 40 years,
three important methodological levels have been (at least in
principle) available to the theoretical chemist pursuing molecular electronic structure: the Hartree-Fock method (9), the
multi-configuration self-consistent field (MCSCF) technique (10),
and large-scale configurational expansions (11-14). Until
recently, most theoretical progress in the field came from inno-
vations in one of these three areas. Moreover, important
advances continue to be made in all three disciplines, and there
is every reason to believe progress will continue. However, a
"second dimension," so to speak, among *ab initio* methodologies
appeared about a decade ago with Pulay's work (15) on analytic
energy derivatives in closed-shell single configuration SCF
theory. Analytic second derivative methods are now available
for nearly all types of Hartree-Fock wavefunctions (16,17) and
also for correlated wavefunctions (18). The derivative methods
can also be used for finding properties such as the dipole
moment and polarizability that are defined as first or second
order changes in the energy with respect to an external perturbation. Now, the reason these analytic derivative methods open
up a new dimension in theoretical chemistry is not so much the
beauty and elegance of the formalism, but rather the inherent
power of the methods. For example, the analytic determination

of the harmonic force constants on a modest sized molecule such
as fluoroformaldehyde (HFC=O) requires perhaps ten times less
effort than determination via finite differences of individual
energies (19). As with the original "dimension" of electronic
structure theory, one can anticipate continued progress in the
development of derivative methods.

The pace of new developments in the theory of electronic
structure has surely become faster with each passing decade. We
see this happening right along with developments in computer
technology already mentioned. As computers improve (in speed,
size, and cost), the sophistication of the computational treatments also seems to improve. Quantum chemical calculations can
now be done on ever larger molecules, with ever more complete
wavefunction determinations, and with ever more chemically useful results, e.g., vibrational frequencies and transition
probabilities. A glance through journals aimed at organic and
inorganic chemists, for instance, shows how wide spread applications studies are. The need is even greater than what the number
of published papers suggests. For this reason, new computer
technology must be rapidly assimilated into theoretical/computa-
tional chemistry in order to provide essential electronic structure data needed throughout the chemical sciences. That means
that there must be greater accessibility to supercomputers and
superminicomputers. The articles in this volume serve as clear
evidence that quantum chemists are quite prepared to use these
systems effectively.

Returning to the estimate of the ideal computing level for
the late 1980's mentioned above and considering the experiences
of the last decade, one is led to the suggestion (actually made
during the course of the workshop discussions) that policies
should be formulated toward dedicating whole supercomputers to
some small number of theoretical/computational chemistry groups.
Dedicated systems continue to offer unassailable advantages,
and in fact (as also brought out during the workshop discussion),
even with supercomputers already in use, some of the milestone
calculations (20) have been done with minicomputers largely
because they were dedicated systems. Until hardware costs have
declined a good bit more, however, it is unlikely that most
research budgets can afford the ideal. In the meantime, the more
affordable dedicated superminicomputer systems should be a more
common instrument among theoretical chemistry groups, or there
should be significant enough time available on supercomputer
systems to justify the added effort of using these more specialized machines. Until this change occurs, a recommendation much
in line with the 1982 Lax Report (21), the chemical sciences will
be deprived of the full resources available from electronic
structure theory.

REFERENCES

1. A. D. McLean, J. Chem. Phys. 32, 1595 (1960).

2. H. F. Schaefer, Proc. IEEE Comp. Sci. Conf., Washington, D.C.
    (1975).

3. H. F. Schaefer and W. H. Miller, Computers and Chemistry
    1, 85 (1976).

4. A. L. Robinson, Science 193, 470 (1976); W. G. Richards,
    Nature 266, 18 (1971).

5. P. K. Pearson, R. R. Lucchese, W. H. Miller and H. F.
    Schaefer, in Minicomputers and Large Scale Computation,
    ACS Symposium Series 57, ed. P. Lykos, pp. 171-190 (Ameri-
    can Chemical Society, Washington, D.C., 1977).

6. Report PB-267-575, National Technical Information Service,
    Department of Commerce, Springfield, Virginia.

7. A. B. Kunz and C. P. Flynn, Int. J. Quant. Chem. Symp. 17,
    574 (1983); A. B. Kunz, ibid. 623.

8. R. A. Bair and T. H. Dunning, Jr., J. Comp. Chem. 5, 44
    (1984).

9. F. W. Bobrowicz and W. A. Goddard, in Vol. 3, "Modern The-
    oretical Chemistry," ed. H. F. Schaefer (Plenum, New York,
    1977).

10. NRCC Proceedings No. 10, Report LBL-12151 (Lawrence Berkeley
    Laboratory, University of California, Berkeley, 1981).

11. I. Shavitt, in Vol. 3, "Modern Theoretical Chemistry," ed.
    H. F. Schaefer (Plenum, New York, 1977).

12. R. J. Bartlett, Ann. Rev. Phys. Chem. 32, 359 (1981).

13. P. G. Jasien and C. E. Dykstra, J. Chem. Phys. 76, 4-64 (1982).

14. I. Shavitt, ACS Subdivision of Theoretical Chemistry News-
    letter, p. 14-20 (1982).

15. P. Pulay, Mol. Phys. 17, 197 (1969); 18, 473 (1970); 21, 329
    (1971).

16. J. A. Pople, R. Krishnan, H. B. Schlegel and J. S. Binkley,
    Int. J. Quant. Chem. Symp. 13, 225 (1979).

17.   Y. Osamura, Y. Yamaguchi, P. Saxe, M. A. Vincent, J. F. Gaw
      and H. F. Schaefer, Chem. Phys. 72, 131 (1982).

18.   D. J. Fox, Y. Osamura, M. R. Hoffmann, J. F. Gaw, G.
      Fitzgerald, Y. Yamaguchi and H. F. Schaefer, Chem. Phys.
      Lett. 102, 17 (1983).

19.   H. B. Schlegel, in "Computational Theoretical Organic
      Chemistry," eds. I. G. Csizmadia and R. Daudel (Reidel,
      Dordrecht, Holland, 1981); Y. Osamura, Y. Yamaguchi and
      H. F. Schaefer, J. Chem. Phys. 77, 383 (1982).

20.   P. Saxe, D. J. Fox, H. F. Schaefer and N. C. Handy, J. Chem.
      Phys. 77, 5584 (1982).

21.   P. D. Lax, "Report of the Panel on Large Scale Computing in
      Science and Engineering" (National Science Foundation,
      Washington, D.C., 1982).

PROBLEM LIMITATIONS AND COST EFFECTIVENESS CONSIDERATIONS IN
COMPUTATIONAL QUANTUM CHEMISTRY

J. S. Binkley

Theoretical Division
Sandia Laboratory
Livermore, California 94550

INTRODUCTION

The computational demands of quantum chemistry have histori-
cally outpaced the capabilities of digital computer systems since
the introduction of such systems nearly forty years ago. This
situation has prevailed despite an essentially astronomic growth
in hardware and software capacities. It demonstrates critical
limitations in solving theoretical chemistry problems and reflects
the clear need for the theoretical chemistry community to obtain
the most cost effective computational resources.

Clearly, there are many individual aspects of both cost
effective computing and quantum chemistry problem limitations that
are specific to each researcher or group of researchers in the
field. These aspects depend in part on the working environment
and availability of computational and manpower resources. The
purpose of this article is to deal briefly with these issues,
i.e., limitations which are presently encountered in quantum
chemistry as well as cost effective computing. The material presented in this article is an outgrowth of a group discussion that
was conducted at the NATO Advanced Research Workshop that is the
topic of this volume. As such, these ideas are not the result of
any single individual.

The next section of this article briefly outlines some of the
limitations currently encountered in solving quantum chemistry
problems. The third section is devoted to some aspects of
cost effective computing. The final section presents discussion
relevant to the impact on quantum chemistry of future growth in
computer architecture.

C. E. Dykstra (ed.),
Advanced Theories and Computational Approaches to the Electronic Structure of Molecules, 203–207.
© 1984 by D. Reidel Publishing Company.

PROBLEM LIMITATIONS ENCOUNTERED IN QUANTUM CHEMISTRY

Problem limitations in quantum chemical research can be divided roughly into two groups: computer limitations and human
limitations. These two areas are treated separately below.
Clearly, these problems arise as a natural consequence of the
mathematical and philosophical tools that are used to solve the
Schroedinger equation. Although the mathematical and philosophical problems require careful treatment, it is beyond the scope of
this study to approach that problem.

1.   Computer Limitations

The computer limitations faced by quantum chemistry researchers
are essentially the same as those faced by computational researchers
in other disciplines. These limitations include the availability
of   computer time and limitations related to computer hardware
and software architectures. The principal limitation is the
availability of high performance computer time. This situation
is most acute in academic institutions where researchers must
either compete for university resources or must maintain their
own dedicated minicomputer facility. In either case, few academic
researchers have ready access to state-of-the-art, high performance
mainframe computer systems (Cray-1S, Cray-XMP, Cyber-207, etc.).
The national laboratories, both in the U.S. and abroad, generally
have such computer systems available for use by quantum chemists.
However, there is considerable competition for these resources
and few research groups can get significant amounts of time on
these systems.

A considerable reduction in the price of high performance
mainframe computers has occurred in the past two years. This
price reduction is coupled to an increased awareness that the
availability of such systems has important implications in computational research which have wide ranging technological benefits.
Thus, there are indications that these computer systems may be
generally more available in the near future than they have been in
the past.

The advent of the 64-bit attached processor has made a dramatic impact on the availability of machine time for quantum
chemistry. A number of research groups (see for example, the
article by Bair and Dunning elsewhere in this volume) have obtained these systems and have successfully implemented quantum
chemistry codes. These systems afford roughly an order of magnitude increase in computer capacity over comparably priced minicomputer systems that are available today. (An additional order
of magnitude increase is possible with large scientific mainframes.)

Many stringent limitations on solving quantum chemistry problems are imposed by the details of computer hardware and software
architectures. The hardware architectural issues of most pressing
concern to theoretical chemists are storage and processing capacity. Quantum chemical calculations require a large capacity to
store and retrieve floating point numbers. Both high-speed (i.e.
central) memory and external storage (e.g. disks) are important
in this regard. The numerous technological advances in recent
years have resulted in computer systems with much larger central
memory units. This has had a considerable impact on the development of algorithms in computational quantum chemistry as well as
in other related fields.

However, it is not likely that it will be possible to perform large calculations in their entirety using central memory
capacities that will be available in the next five to ten years.
Thus, there will be continued reliance on external storage devices.
Although there have been significant recent advances in the processing speed of central and attached processing units, the advances in large capacity, high speed external storage units has
not been as dramatic. In most large scientific applications, the
processing speed of the central processing unit far exceeds the
ability of the input/output system to move data rapidly enough.
To illustrate this point, consider the following. At peak performance, a Cray-1S can perform floating point operations approximately 400 times faster than a VAX. The disk units on a Cray
system can transfer data approximately four times faster than the
VAX disks. As a result, it is extremely important to carefully
optimize both the floating point arithmetic *and* input/output portions of a quantum chemistry algorithm.

Limitations related to computer software are imposed by the
degree of difficulty in developing algorithms as well as the
transportability and interchangeability of software between research groups. Both of these areas are becoming more troublesome
as hardware architectures become more exotic. Computer code
development problems encountered in vector oriented systems such
as the Cray are similar to those encountered in pipeline oriented
attached processors. However, there are enough differences so
as to frequently make codes not interchangeable between these
systems. Many of the problems arise from differences between the
exact details of operating system calls in any given environment.
However some problems of this nature can be overcome by developing
software packages in separate environments that perform vital
functions in a fully consistent manner.

2.   Human Limitations

Progress in many theoretical chemistry groups is dependent
on the efforts of a relatively small number of highly qualified

individuals. These individuals perform a number of complicated
tasks comprising the development of new methods, design of computer algorithms, computer code development as well as pursuing
chemical applications of the new methods.

There has been a tendency in recent years for computer systems to become considerably more complex. This has resulted in
increased human costs in developing new quantum chemical methods
as well as consuming considerable time in optimizing the computer
implementations of older methods. Frequently, the new architectures require total recoding of existing algorithms. The incen-
tive in this process is usually reduced computing costs. However,
in some cases, the improved performance obtained by rewriting the
computer codes enables researchers to solve fundamentally new
problems.

The fundamental manpower limitations are acutely felt in both
academic and national laboratory environments. This is due to
the fact that there has been considerable turnover in hardware
systems in both areas. The introduction of large vector oriented
mainframes as well as the smaller pipeline oriented attached processors is producing an extensive set of new quantum chemistry
applications programs. Again, recourse to well defined, efficiently
programmed library utilities for performing frequently used operations is vital to the timely development of these new codes. Furthermore, this avenue offers some hope in dealing with forthcoming
architectural developments.

COST EFFECTIVENESS CONSIDERATIONS IN COMPUTATIONAL QUANTUM CHEMISTRY

The definition of what constitutes cost effective computing
is specific to any given research group. In general, cost effectiveness considerations require careful optimization of a number
of vital resources which include (but are not limited to) availability of computer time, manpower availability, and software
availability. These areas are usually interrelated in a complicated manner and it is therefore difficult if not impossible to
make any broad sweeping generalizations. Issues related to the
availability of computer time and manpower resources have been
dealt with above. However, there are several other factors which
may be considered.

The cost effectiveness of computing in computational quantum
chemistry is frequently a moot point. A researcher may have ready
access to only one form of computational service. Only in rare
cases do the specific needs of a quantum chemistry research group
give rise to the computer hardware that is actually made available.
Most researchers have access to resources that are shared at
group or department level or resources that are shared within a

laboratory or at the national level. In this sense, the only
relevant cost effectiveness considerations are those related
to obtaining the maximum performance in the local environment.
All other considerations become secondary.

Frequently, compatibility with existing hardware and/or software features is an important issue. Many researchers have used
similar codes for an extended period of time. As a result, compatibility with previously established methods is an important
concern.

Local vs. remote computing resources is another issue that
is of importance to many research groups. The decentralization
of computing facilities that has resulted from the supply of
moderately priced minicomputer systems has greatly increased
the ability of research groups to exert more control of their
destinies.

In summary, cost effectiveness considerations almost never
reduce to studying the cost of basic computer operations. Other
factors such as those described above usually play a more important role than simple cost considerations.

CONCLUSIONS

This article has presented discussion related to some of the
limitations encountered in solving quantum chemistry problems as
well as the relationship of these limitations to obtaining cost
effective computing. Most of the issues dealt with in this
article are well known to researchers that are actively involved
in this area of science.

It is clear that the cycle of hardware upgrade followed by
re-development of computer algorithms will continue to be a per-
sistent problem in computational quantum chemistry. If anything,
it will become considerably more acute in the next few years as
multi-processing computers become available. Hopefully, the pipeline and vector oriented features that facilitate high processing
rates in single processor systems will continue to be present in
these newer systems.

It is imperative for computational quantum chemistry to come
to grips with the continued re-development cycle. One avenue that
offers considerable hope is to develop computer algorithms using
generally available libraries of commonly performed computer
operations.

ALGORITHMIC CONSIDERATIONS IN LARGE MAINFRAME COMPUTERS

J. S. Binkley

Sandia National Laboratories
Livermore, CA 94550

1. Introduction

Computational quantum chemistry has traditionally been at the forefront of scientific applications exploiting the rapid growth of computational capabilities. This state of affairs has resulted from the complexity of the problems that computational quantum chemists desire to solve. As a result of the basic formalism of quantum mechanics as expressed in matrix form, this field of science is acutely dependent on the capability of performing high-speed floating point arithmetic as well as being able to manipulate large quantities of data.

Several important advances in computational technology have occurred in the last decade or two which have been of extreme interest to quantum chemists. The first was the development of large, very fast, general purpose mainframe computers. For many years, this form of computing machine was the mainstay of quantum chemistry. Somewhat later came the introduction of special purpose attached processors. Although such systems were generally available almost ten years ago, few quantum chemists took advantage of their capabilities due to their limited numerical precision.

More recently (since approximately 1976), advances have occurred in both of these areas. Also, during this period, it has become possible to procure super-minicomputers that have computational capabilities comparable to the large mainframe computers of a decade ago. The fastest mainframe computers available today have cycle times under ten nanoseconds and have been augmented with special hardware that premits extremely

C. E. Dykstra (ed.),
Advanced Theories and Computational Approaches to the Electronic Structure of Molecules, 209–216.
© *1984 by D. Reidel Publishing Company.*

efficient computations on vectors of integers and floating point numbers. The attached processors, which employ a pipeline architecture, have been improved and are now capable of dealing with floating point numbers with sufficient precision for quantum chemistry.

These advances in computational technology have not been without a price for quantum chemists. The newer computing machines (both vector- oriented mainframes and pipeline-oriented attached processors) require that calculational data be organized in fundamentally different ways than before. After yesrs of dealing with computer systems that were oriented towards efficient processing of scalar data, quantum chemistry has arrived at a point where scalar algorithms and scalar thinking are very entrenched. In order to most effectively utilize the newer architectures it has been necessary to develop new computer algorithms.

In some areas, particularly self-consistent field (SCF) algorithms, configuration interaction (CI) algorithms and many-body perturbation theory (MBPT) algorithms, the process of re-developing algorithms has been relatively straightforward. This results from the facts that these forms of computations are originally based on linear algebra operations. Other steps encountered in quantum chemistry calculations, notably the evaluation of one- and two-electron integrals over Gaussian basis functions, are significantly more difficult to re-cast for vector and pipeline computers.

In either case, for both the easy-to-convert and hard-to-convert steps, there has been considerable upheaval. Even for computational steps that are easily cast into matrix form (SCF, CI and MBPT) there are always other considerations such as optimization of input/output (I/O) and efficient utilization of high-speed memory that may necessitate re-design of existing algorithms. Thus, for both these and the hard-to-convert algorithms, complete re-coding is frequently necessary. Another feature that is often encountered when dealing with vector and pipeline computing machines is that one must often resort assembly language (i.e. machine coding) to fully utilize the unique hardware features.

The driving force that justifies all of this work should be clear. The high-performance pipelining and vector computers that are available today permit processing rates that are from ten to 1000 times faster than the super-minicomputers that have become the mainstay of computational quantum chemistry since about 1978. This increased level of performance translates directly into new chemistry problems that can be solved with existing methodology as well as providing a more fertile area for developing more powerful methods.

The purpose of this article is to list and briefly describe some techniques that the author has found useful in developing algoritms for vector-oriented large mainframe computers. Some attention will also be given th problems that occur frequently in developing vectorized algorithms. The strategy of this article is to deal with this problem in as general a manner as possible. Specific examples of vectorized and pipelined algorithms for the computational steps encountered in quantum chemistry calculations are given elsewhere in this volume.

The next section introduces concepts which are important in understanding vectorized algorithms. The third section lists some frequently encountered problems and some frequently (but sometimes unrelated) solutions. Finally, concluding remarks are presented.

Algorithmic Concepts

One of the most important concepts in developing vectorized algorithms is that there are competing resources in a large mainframe computer that must be balanced in a (usually) complex manner. In most cases, an optimum algorithm must minimize the total computer resources consumerd in a given calculation. Frequently, the definition of these resources and their relative importance is embodied in the particular computer billing algorithm that is used. Thus, any discussion of specifics is necessarily complicated. In this article, we consider that the main resources to be minimized comprise central memory, central processing time (CP-time), input/outpur time (I/O time) and time integrated utilization of external storage. Note that in the case of I/O time, both the number of I/O requests and the number of computer words transferred are important.

Modern, high-speed vector processing machines have some idiosynchrasies that most developers of scalar algorithms have not previously encountered. For example, although these machines have impressive capabilities in performing floating point operations, their I/O capabilities employ peripheral devices that are comparable to those available on super minicomputers. Thus, algorithms that were developed on older machines where CP-time and I/O times were well balanced become terribly I/O bound.

Another idiosynchrasy of these machines is that logical operations decision making processes and memory references frequently take more CP-time than do floating-point operations. As a result, it is necessary to organize most computational

steps so as to minimize the amount of decision making. In some
cases, it is advantageous to perform extra arithmetic operations
on portions of data to avoid unnecessary decision making. This
situation is completely different from that encountered in
scalar machines where logical decisions can usually be made in
less time than floating point operations.

These two features (i.e. relatively slow I/O and long
decision making times) lead to the important conclusion that
data structures play an important role in developing algorithms
for vector machines. Thus, algorithmic development in these
machines is often drives the search for more effective ways in
which to organize the data in preparation for processing.
Algorithms which require excessive decision making or memory
referneces will generally run slower than those algorithms which
minimize those items.

A frequent outcome of considering these features is the
production of better scalar algorithms. Although most scalar
computers are not as sensitive to memory references and logical
decision making, reductions in this area almost always lead to
concommitant reductions in execution time. A notable exception
to this occurs in those algorithms where the imposition of a
data structure leads to many redundant arithmetic operations.

Objective testing of algorithms is the last general concept
to be dealt with in this section. In the course of solving
computational problems, it is frequently possible to design
several different algorithms. In general, any one of these
multiple algorithms will be more effective in utilizing a
particular machine resource than will be the others. However,
the overall performance of an algorithm must be measured as a
function of total system utilization of resources. An important
outcome of this is that it is necessary to quantitatively
evaluate relative algorithms. In most cases this implies that
the algorithm must be fully developed, coded and debugged. In
many cases, it is nearly impossible to predict accurately which
algorithm will be the best.

To conclude this section, we summarize the three main
points that were covered. First, data structures play a
fundamental role in the shaping of vector an pipeline
algorithms. Second, the search for an algorithm that most
effectively minimizes competing machine resources frequently
leads to better scalar algorithms. Finally, there exists an
acute necessity to objectively and quantitatively evaluate
competing algorithms.

Problems and Solutions

This section presents a few of the problems that arise when developing algorithms for a vector oriented machine along with solutions that have found to be useful. The treatment here is brief and by no means comprehensive. Exhaustive examples of detailed algorithms are presented and discussed elsewhere in this volume.

As described above, excessive I/O times are a serious problem on these computing machines. This results from the fact that the I/O subsystems present on large vector mainframes are very slow compared to the central processing units. Although solid-state disks are slowly becoming available, they are still relatively costly. Hence, effective optimization of the I/O portions of quantum chemistry codes is a vital necessity. A number of techniques are of general interest. First, I/O times can be reduced by simply reducing the amount of I/O through use of greater amounts of central memory. Many scientific applications have total execution times that are a linear function of the amount of memory that is available to store calculational data. Thus, by re-dimensioning or by dynamic allocation of memory, it is possible to reduce the I/O costs associated with any give calculation. Obviously, the savings as a function of total system resources depend on how critical memory resources are.

Another approach to reducing the amount of I/O time in a calculation is to overlap the I/O time with either processing of the data or with other I/O. Most large mainframes have software that permits the user to perform I/O asynchronously with computing. For example, in Hartree-Fock SCF calculations, it is reasonable to be processing the data from one record of the two-electron integral file while the next one is being read into central memory from the disk. This technique of double-buffering can be employed in many applications and is quite effective in reducing the total amount of system time a calculation will consume. Overlapping I/O with other I/O is a related but less frequently used method for reducing total I/O times. This form of optimization takes advantage of the fact that the I/O hardware on most large scientific computers is organized so that there are several channels into memory from several sets of disks. Thus, by splitting a large file up into pieces which reside on separate I/O channels, it is possible to read the total file in much less elapsed time.

The concept of data structures was introduced in the previous section. The use of new data structures is probably the most powerful technique in developing algorithms for vector processing machines. Perhaps the simplest example of this

technique is sorting the two-electron integrals into square
canonical order. In this manner, the integral file can be
processed in matrix form. This technique has the disadvantage
that zero matrix elements are explicitly stored thereby
increasing the required amount of disk storage (and also the I/O
time). However, even though there are more integrals to deal
with in any subsequent step of the calculation, it is usually
possible to achieve maximum computing rates. Thus, if at least
one tenth of the integrals are non-zero, increased efficiency is
obtained. Clearly, any reduction in execution time that is
realized must be measured against the time required to sort the
integrals and the additional I/O time that results from storing
and retrieving a larger data structure.

Through the use of GATHER/SCATTER operations, it is
possible to avoid the introduction of unnecessary zero data
elements in a data structure. For example, if the non-zero
elements of a vector are stored along with indexes that specify
their ordinal positions it is possible to use the indices to
extract the corresponding elements from another vector to
perform a scalar product. Thus, one uses the GATHER function to
collect together the desired elements. The SCATTER function is
used to perform the inverse function. Most vector mainframes
implement these operations in software. Thus, this technique is
most advantageous in the limit of few non-zero elements.

A technique of handling data structures that is directly
related to GATHER/SCATTER is the use of bit-maps. In this
method, a string of bits that is the length (in bits) of the
total data structure is used to indicate which elements are
present (bit set to one) and which are absent (bit set to zero).
Relative to GATHER/SCATTER, this technique has the advantage of
reducing the number of computer words occupied by the data
structure since the number of indices is reduced by a factor of
the number of bits in a computer word. The disadvantage of this
technique is that it usually requires functions that operate on
the bit map to be implemented in assembly language. This
disadvantage is partially offset by the fact that most large
mainframe computers rely on bit strings for their internal
operating system functions and therefore have highly specialized
instructions for manipulating bit strings.

As a practical example of the data structure techniques
cited above we consider the processing of the two-electron
intregrals in the Hartree-Fock SCF step (i.e., full storage,
partial storage with GATHER/SCATTER, and partial storage with
bit maps). Assuming that the original time required for an SCF
iteration was one hundred units of CP-time, the times after
introducing these types of data structures are approximately
one unit, 33 units and 25 units of time, respectively. The

analogous I/O times (starting with 100 units of I/O time) are
approximately 200, 100, and 50 units of I/O time, respectively.
Note that the I/O times do not include time required to sort the
integrals into canonical order. Clearly, selection of which
algorithm to be used is not straightforward. In the presence of
other computer users that have CP intensive application but do
little I/O, the first algorithm leads to the most effective
utilization of resources. In a "stand-alone" environment (e.g.
on a dedicated attached processor, the last algorithm may be
most effective.

Conclusions

Several concepts which are important in the development
of vector algorithms have been presented and discussed along
with examples of problems which are encountered in the
development of such algorithms. The creation of new algorithms
for vector and pipeline machines is a tedious and frustrating
process due the the necessity to quantitatively evaluate each
variant of each algorithm. This reflects the fact that the
architectures of these machines are more complicated than scalar
computing machines in complex and usually subtle ways. Thus
algorithms depend on the exact organization of the calculational
data that is to be processed.

As can be seen from the articles in this volume, various
vector and pipeline computers have been available to quantum
chemists in recent years. The talents of numerous scientists
have been devoted to the time intensive effort of reorganizing
quantum chemistry codes to effectively use these new computing
machines. Although the revised codes have made it possible to
apply existing computational methods to larger, more complex
chemistry problems, the conversion effort has diverted many of
the same people who formerly developed new methods. Thus, the
overall benefit to theoretical chemistry has been somewhat
mixed.

One action that will maximize the efforts of these people
is to capture their efforts in software libraries that are more
generally accessible. To date, efforst in this area have met
with only limited success. However, in the future, the
availability of such libraries may make a critical difference in
the impact of quantum chemistry in solving problems of
technological importance.

Complicating this is the fact that the next generation of
computing hardware is already taking the form of parallel
processing machines. In a short time, quantum chemistry may be
faced with another major upheaval in computing software when it

becomes necessary to devise wholly new algorithms. In the
coming generation, communications problems will probably become
the dominant factors in the evolution of algorithms. Hopefully,
the new architectures will still employ hardware that has vector
and pipeline features thereby extending the life of algorithms
that are being developed today.

BIBLIOGRAPHY

The following list of papers was prepared from suggestions of the participants. Not a comprehensive list, it is intended to provide background resources for topics discussed at the workshop and, in particular, the papers presented by the participants.

UTILIZING VECTOR AND ARRAY PROCESSORS

1. "The Architecture of Pipelined Computers" P. M. Kogge (McGraw-Hill, New York, 1981).

2. "Parallel Computations" G. Rodrique, ed. (Academic Press, New York, 1982).

3. "A Vectorizable Eigenvalue Solver for Sparse Matrices" L. C. Bernard and F. J. Helton, Comp. Phys. Commun. $\underline{25}$, 73 (1982).

4. "Application of the Cray-1 for Quantum Chemistry Calculations" V. R. Saunders and M. F. Guest, Comp. Phys. Commun. $\underline{26}$, 389 (1982).

5. "Quantum Chemistry with an Attached Processor" R. A. Bair and T. H. Dunning, Jr., J. Comp. Chem. $\underline{5}$, 44 (1984).

6. "Quantum Chemical Calculations Using an FPS-164 Attached R. Shepard, R. A. Bair, R. A. Eades, A. F. Wagner, L. B. Harding, M. J. Davis, and T. H. Dunning, Jr., Int. J. Quantum Chem. $\underline{S17}$, 613 (1983).

7. "Supercomputers in Chemistry" P. Lykos, ed., Am. Chem. Soc. Symp. Ser. 173 (ACS, Washington, D.C., 1981).

8. "Supercomputers = Colossal Computations + Enormous Expectations + Renowned Risk" N. R. Lincoln, IEEE Computer $\underline{16}$, 38 (1983).

9. "A Series of Tests of Small- and Medium-Scale Computers
    Commonly Used for Computations by Solid-State Theorists and
    Quantum Chemists" A. B. Kunz, Int. J. Quantum Chem. S17,
    623 (1983).

LARGE SCALE CI METHODOLOGY

10. "Group Theoretical Approach to the Configuration Interaction
    and Perturbation Theory Calculations for Atomic and Molecular
    Systems" J. Paldus, J. Chem. Phys. 61, 5321 (1974).

11. "Many-Electron Correlation Problem. A Group Theoretical
    Approach," J. Paldus, in Vol. 2, "Theoretical Chemistry.
    Advances and Perspectives" (Academic Press, New York, 1976).

12. "Matrix Element Evaluation in the Unitary Group Approach to
    the Electron Correlation Problem" I. Shavitt, Int. J. Quantum
    Chem. S12, 5 (1978).

13. "The Loop-Driven Graphical Unitary Group Approach: A Power-
    ful Method for the Variational Description of Electron Cor-
    relation" B. R. Brooks, W. D. Laidig, P. Saxe, N. C. Handy
    and H. F. Schaefer, Phys. Scripta 21, 312 (1980).

14. "Generalizations of the Direct CI Method Based on the Graphi-
    cal Unitary Group Approach. II. Single and Double Replace-
    ments from any Set of Reference Configurations" P.E.M.
    Siegbahn, J. Chem. Phys. 72, 1647 (1980).

15. "Unitary Group Approach to the Many-Electron Correlation
    Problem via Graphical Methods of Spin Algebras" J. Paldus
    and M. J. Boyle, Phys. Scripta 21, 295 (1980).

16. "The Unitary Group for the Evaluation of Electronic Energy
    Matrix Elements" J. Hinze, ed., Vo. 22 of "Lectures Notes
    in Chemistry" (Springer-Verlag, Berlin, 1981).

17. "New Implementation of the Graphical Unitary Group Approach
    for Multireference Direct Configuration Interaction Calcula-
    tions" H. Lischka, R. Shepard, F. B. Brown and I. Shavitt,
    Int. J. Quantum Chem. S15, 91 (1981).

18. "The ALCHEMY Configuration Interaction Method. I. The Sym-
    bolic Matrix Method for Determining Elements of Matrix
    Operators" B. Liu and M. Yoshimine, J. Chem. Phys. 74, 612
    (1981).

19. "The Shape-Driven Graphical Unitary Group Approach to the
     Electron Correlation Problem. Application to the Ethylene
     Molecule" P. Saxe, D. J. Fox, H. F. Schaefer and N. C. Handy,
     J. Chem. Phys. 77, 5584 (1982).

20. "The Direct CI Method. A Detailed Analysis" V. R. Saunders
     and J. H. van Lenthe, Mol. Phys. 48, 923 (1983).

21. "The Unitary Group and the Electron Correlation Problem"
     I. Shavitt in "New Horizons of Quantum Chemistry," eds.
     P.-O. Lowdin and B. Pullman (Reidel, Dordrecht, Holland,
     1983).

MATRIX-FORMULATED CORRELATION METHODS

22. "Theory of Self Consistent Electron Pairs. An Iterative
     Method for Correlated Many-Electron Wavefunctions" W. Meyer,
     J. Chem. Phys. 64, 2901 (1976).

23. "A Theory of Self-Consistent Electron Pairs. Computational
     Methods and Preliminary Applications" C. E. Dykstra, H. F.
     Schaefer and W. Meyer, J. Chem. Phys. 65, 2740 (1976).

24. "Many Body Perturbation Calculations and Coupled Electron
     Pair Models" R. Ahlrichs, Comp. Phys. Commun. 17, 31 (1979).

25. "CEPA Model and MBPT" R. Ahlrichs and C. Zirz, in "Molecular
     Physics and Quantum Chemistry Workshop" (Wollongong, 1980).

26. "Recent Computational Developments with the Self-Consistent
     Electron Pairs Method and Application to the Stability of
     Glycine Conformers" C. E. Dykstra, M. D. Garrett and R. A.
     Chiles, J. Comp. Chem. 2, 266 (1981).

27. "The Self Consistent Electron Pairs Method for Multiconfi-
     guration Reference State Functions" H.-J. Werner and E.-A.
     Reinsch, J. Chem. Phys. 76, 3144 (1982).

28. "An Electron Pair Operator Approach to Coupled Cluster Wave-
     functions. Applications to He_2, Be_2 and Mg_2 and Comparison
     with CEPA Methods" R. A. Chiles and C. E. Dykstra, J. Chem.
     Phys. 74, 4544 (1981).

29. "The Most Efficacious One-Electron Bases for Determining and
     Representing Correlated Molecular Electronic Wave Functions.
     Unity in Seemingly Disparate Electron Correlation Methods"
     P. G. Jasien and C. E. Dykstra, J. Chem. Phys. 76, 4564
     (1982).

(Also: 18,20)

EVALUATION AND PROCESSING OF INTEGRALS

30.   "Gaussian Quadrature Formulas" A. H. Stroud and D. Secrest
      (Prentice Hall, Englewood Cliffs, New Jersey, 1966).

31.   "Numerical Integration Using Rys Polynomials" H. F. King
      and M. Dupuis, J. Comp. Phys. $\underline{21}$, 144 (1976).

32.   "One- and Two-Electron Integrals over Cartesian Gaussian
      Functions" L. E. MacMurchie and E. R. Davidson, J. Comp.
      Phys. $\underline{26}$, 218 (1978).

33.   "Computation of Electron Repulsion Integrals Using the Rys
      Quadrature Method" J. Rys, M. Dupuis and H. F. King, J. Comp.
      Chem. $\underline{4}$, 154 (1983).

34.   "Integral Evaluation Algorithms and Their Implementation"
      D. Hegarty and G. van der Velde, Int. J. Quantum Chem. $\underline{23}$,
      1135 (1983).

COUPLED CLUSTER AND RELATED APPROACHES

35.   "On the Correlation Problem in Atomic and Molecular Systems.
      Calculation of the Wavefunction Components in Ursell-type
      Expansion Using Quantum-Field Theoretical Methods" J. Cizek,
      J. Chem. Phys. $\underline{45}$, 4256 (1966).

36.   "On the Use of the Cluster Expansion and the Technique of
      Diagrams in Calculations of Correlation Effects in Atoms and
      Molecules" J. Cizek, Adv. Chem. Phys. $\underline{14}$, 35 (1968).

37.   "Correlation Problems in Atomic and Molecular Systems. III.
      Rederivation of the Coupled-Pair Many-Electron Using the
      Traditional Quantum Chemical Methods" J. Cizek and J. Paldus,
      Int. J. Quantum Chem. $\underline{5}$, 359 (1971).

38.   "Correlation Problems in Atomic and Molecular Systems. IV.
      Extended Coupled-Pair Many-Electron Theory and Its Applica-
      tion to the BH_3 Molecule" J. Paldus, J. Cizek and I. Shavitt,
      Phys. Rev. $\underline{A5}$, 50 (1972).

39.   "Electron Correlation in Small Molecules" A. C. Hurley
      (Academic Press, New York, 1976).

40.   "Correlation Problems in Atomic and Molecular Systems. V.
      Spin-adapted Coupled Cluster Many-Electron Theory" J. Paldus,
      J. Chem. Phys. 67, 303 (1977).

41.   "Electron Correlation Theories and Their Application to the
      Study of Simple Reaction Poten-ial Surfaces" J. A. Pople,
      R. Krishnan, H, B. Schlegel and J. S. Binkley, Int. J. Quan-
      tum Chem. 14, 545 (1978).

42.   "Many-Body Perturbation Theory, Coupled-Pair Many-Electron
      Theory, and the Importance of Quadruple Excitations for the
      Correlation Problem" R. J. Bartlett and G. D. Purvis, Int.
      J. Quantum Chem. 14, 561 (1978).

43.   "A Coupled-Cluster Approach to the Many-Body Perturbation
      Theory for Open-Shell Systems" I. Lindgren, Int. J. Quantum
      Chem. S12, 33 (1978).

44.   "The Quartic Force Field of H_2O Determined by Many-Body
      Methods that Include Quadruple Excitation Effects" R. J.
      Bartlett, I. Shavitt and G. D. Purvis, J. Chem. Phys. 71,
      281 (1979).

45.   "Orthogonally Spin-Adapted Coupled Cluster Theory for Closed
      Shell Systems Including Triexcited Clusters" B. G. Adams and
      J. Paldus, Phys. Rev. A20, 1 (1979).

46.   "Coupled Cluster Approach" J. Cizek and J. Paldus, Phys.
      Scripta 21, 251 (1980).

47.   "Molecular Applications of Coupled Cluster and Many-Body
      Perturbation Methods" R. J. Bartlett and G. D. Purvis,
      Physica Scripta 21, 255 (1980).

48.   "Contribution of Triple Substitutions to the Electron Cor-
      relation Energy in Fourth Order Perturbation Theory" R.
      Krishnan, M. J. Frisch and J. A. Pople, J. Chem. Phys. 72,
      4244 (1980).

49.   "Atomic Many-Body Theory" I. Lindgren and J. Morrison
      (Springer, New York, 1981).

50.   "Coupled Cluster Method for Multideterminental Reference
      States" B. Heziorski and H. J. Monkhorst, Phys. Rev. A242,
      1668 (1981).

51.   "Many-Body Perturbation Theory and Coupled-Cluster Theory
      for Electron Correlation in Molecules" R. J. Bartlett, Ann.
      Rev. Phys. Chem. 32, 359 (1981).

52. "An Efficient and Accurate Approximation to Double Substitution Coupled Cluster Wavefunctions" R. A. Chiles and C. E. Dykstra, Chem. Phys. Lett. 80, 69 (1981).

53. "A Full Coupled-Cluster Singles and Doubles Model: The Inclusion of Disconnected Triples" G. D. Purvis and R. J. Bartlett, J. Chem. Phys. 76, 1910 (1982).

54. "Coupled Cluster Approach in the Electronic Structure Theory of Molecules" V. Kvasnicka, V. Laurinc, S. Biskupic and M. Haring, Adv. Chem. Phys. 52, 181 (1983).

(Also: 28)

MULTI-CONFIGURATION REFERENCE WAVEFUNCTIONS

55. "Improved Quantum Theory of Many-Electron Systems" W. A. Goddard III, Phys. Rev. 157, 81 (1967).

56. "Configuration Interaction Studies of Ground and Excited States of Polyatomic Molecules. I. The CI Formulation and Studies of Formaldehyde" J. L. Whitten and M. Hackmeyer, J. Chem. Phys. 51, 5584 (1969).

57. "Classification of Configurations and the Determination of Interacting and Noninteracting Spaces in Configuration Interaction" A. D. McLean and B. Liu, J. Chem. Phys. 58, 1066 (1973).

58. "Theoretical Studies of Curve Crossing: Ab Initio Calculations on the Four Lowest $^1\Sigma^+$ States of LiF" L. R. Kahn, P. J. Hay and I. Shavitt, J. Chem. Phys. 61, 3530 (1974).

59. "The Generalized Valence Bond Description of O_2" B. J. Moss, F. W. Bobrowicz and W. A. Goddard II, J. Chem. Phys. 63, 4632 (1975).

60. "Generalized Valence Bond Calculations on the Ground State of Nitrogen" T. H. Dunning, Jr., D. C. Cartwright, W. J. Hunt, P. J. Hay and F. W. Bobrowicz, J. Chem. Phys. 64, 4755 (1976).

61. "The Low-Lying States of Hydrogen Fluoride: Potential Energy Curves for the $X^1\Sigma^+$, $^3\Sigma^+$, $^3\pi$ and $^1\pi$ States" T. H. Dunning, Jr. J. Chem. Phys. 65, 3854 (1976).

62. "MCSCF Studies of Chemical Reactions: Natural Reaction
     Orbitals and Localized Reaction Orbitals" K. Ruedenberg and
     K. R. Sundberg, in "Quantum Science," eds. J.-L. Calais,
     O. Goscinski, J. Linderberg and Y. Ohrn (Plenum, New York,
     1976).

63. "The Method of Configuration Interaction" I. Shavitt in
     Vol. 3 "Modern Theoretical Chemistry," ed. H. F. Schaefer
     (Plenum, New York, 1977).

64. "Applicability of the Multi-reference Double-Excitation CI
     (MRD-CI) Method to the Calculation of Electronic Wavefunc-
     tions and Comparison with Related Techniques" R. J. Buenker,
     S. D. Peyerimhoff and W. Butscher, Mol. Phys. $\underline{35}$, 771 (1978).

65. "Optimization of Orbitals for Multiconfiguration Reference
     States" E. Dalgaard and P. Jorgensen, J. Chem. Phys. $\underline{69}$,
     3833 (1978).

66. "Convergency Studies of Second and Approximate Second Order
     Multiconfiguration Hartree-Fock Procedures" D. L. Yeager
     and P. Jorgensen, J. Chem. Phys. $\underline{71}$, 755 (1979).

67. "Electronic Rearrangements During Chemical Reactions. II.
     Planar Dissociation of Ethylene" L. M. Cheung, K. R. Sundberg
     and K. Ruedenberg, Int. J. Quantum Chem. $\underline{16}$, 1103 (1979).

68. "A Complete Active Space SCF Method (CASSCF) Using a Density
     Matrix Formulated Super-CI Approach" B. O. Roos, P. R. Taylor
     and P.E.M. Siegbahn, Chem. Phys. $\underline{48}$, 157 (1980).

69. "The Complete Active Space SCF Method in a Fock-Matrix-Based
     Super-CI Formulation" B. O. Roos, Int. J. Quantum Chem. $\underline{S14}$,
     175 (1980).

70. "A Quadratically Convergent Multiconfiguration Self-Consistent
     Field Method with Simultaneous Optimization of Orbitals and
     CI Coefficients" H.-J. Werner and W. Meyer, J. Chem. Phys.
     $\underline{73}$, 2342 (1980).

71. "A Quadratically Convergent MCSCF Method for Simultaneous
     Optimization of Several States" H.-J. Werner and W. Meyer,
     J. Chem. Phys. $\underline{74}$, 5794 (1981).

72. "A Second Order MCSCF Method for Large CI Expansions" B.
     Lengsfield and B. Liu, J. Chem. Phys. $\underline{75}$, 478 (1981).

73. "The Coupled-Cluster Method with a Multiconfiguration Refer-
     ence State" A. Banerjee and J. Simons, Int. J. Quantum Chem.
     $\underline{19}$, 207 (1981).

74. "Coupled-Cluster Method for Multideterminantal Reference
     States" B. Jeziorski and H. J. Monkhorst, Phys. Rev. A24,
     1668 (1981).

75. "Are Atoms Intrinsic to Molecular Electronic Wavefunctions?
     I. The FORS Model" K. Ruedenberg, M. W. Schmidt, M. M.
     Gilbert and S. T. Elbert, Chem. Phys. 71, 41 (1982).

76. "General Second-Order MCSCF Theory for Large CI Expansions"
     B. H. Lengsfield, J. Chem. Phys. 77, 4073 (1982).

77. "Proper Characterization of a Multiconfiguration Self-Con-
     sistent Field State" J. T. Golab, D. L. Yeager and P.
     Jorgensen, Chem. Phys. 78, 175 (1983).

78. "Optimization and Characterization of a Multiconfigurational
     Self-Consistent Field (MCSCF) State" J. Olsen, D. L. Yeager
     and P. Jorgensen, Adv. Chem. Phys. 54, 1 (1983).

(Also: 27)

ENERGY DERIVATIVES AND MOLECULAR PROPERTIES

79. "Direct Use of the Gradient for Investigating Molecular
     Energy Surfaces" P. Pulay, in Vol. 4 "Modern Theoretical
     Chemistry", H. F. Schaefer, ed., (Plenum, New York, 1977).

80. "Ab Initio Theory of the Polarizability Derivatives in Hydro-
     gen Sulfide" R. L. Martin, E. R. Davidson and D. F. Eggers,
     Chem. Phys. 38, 3111 (1979).

81. "Derivative Studies in Hartree-Fock and Moeller-Plesset
     Theories" J. A. Pople, R. Krishnan, H. B. Schlegel and J. S.
     Binkley, Int. J. Quantum Chem. S13, 225 (1979).

82. "Derivative Studies in Configuration Interaction Theory"
     R. Krishnan, H. B. Schelgel and J. A. Pople, J. Chem. Phys.
     72, 4654 (1980).

83. "Analytic Configuration Interaction (CI) Gradient Techniques
     for Potential Energy Hypersurfaces. A Method for Open-Shell
     Molecular Wavefunctions" Y. Osamura, Y. Yamaguchi and H. F.
     Schaefer, J. Chem. Phys. 75, 2919 (1981).

84. "Calculation of One-Electron Properties Using Limited Con-
     figuration Interaction Techniques" K. Raghavachari and J. A.
     Pople, Int. J. Quantum Chem. 20, 1067 (1981).

85. "Molecular Orbital Studies of Vibrational Frequencies" J. A.
    Pople, H. B. Schlegel, R. Krishnan, D. J. Degrees, J. S.
    Binkley, M. J. Frisch, R. A. Whiteside, R. F. Hout and W. J.
    Hehre, Int. J. Quantum Chem. S15, 269 (1981).

86. "Generalization of Analytic Configuration Interaction (CI)
    Gradient Techniques for Potential Energy Hypersurfaces,
    Indluding a Solution to the Coupled Perturbed Hartree-Fock
    Equations for Multiconfigurational SCF Molecular Wavefunc-
    tions" Y. Osamura, Y. Yamaguchi and H. F. schaefer, J. Chem.
    Phys. 77, 383 (1982).

87. "Quadratic Response Functions Within the Time-Dependent
    Hartree-Fock Approximation" E. Dalgaard, Phys. Rev. A26,
    42, (1982).

88. "Multiconfigurational Time-Dependent Hartree-Fock Response
    Functions" P. Jorgensen, P. Swanstrom, D. L. Yeager and
    J. Olsen, Int. J. Quantum Chem. 23, 959 (1983).

89. "Ab Initio Analytical Gradients and Hessians" P. Jorgensen
    and J. Simons, J. Chem. Phys. 79, 334 (1983).

90. "Analytic Energy Second Derivatives for General Correlated
    Wavefunctions, Including a Solution of the First-Order
    Coupled Perturbed Configuration Interaction Equations" D. J.
    Fox, Y. Osamura, M. R. Hoffmann, J. F. Gaw, G. Fitzgerald,
    Y. Yamaguchi and H. F. Schaefer, Chem. Phys. Lett. 102, 17
    (1983).

91. "Molecular Properties from MCSCF-SCEP Wavefunctions. I.
    Accurate Dipole Moment Functions of OH, OH- and OH+"
    H.-J. Werner, P. Rosmus and E.-A. Reinsch, J. Chem. Phys.
    79, 905 (1983).

GENERAL AREAS AND NEW DIRECTIONS

92. "Three Approaches to Electron Correlation" O. Sinanoglu and
    K. A. Bureckner (Yale Press, 1970).

93. "Theoretical Approach to the Calculation of Energies and
    Widths of Resonance (Autoionizing) States in Many-Electron
    Atoms" C. A. Nicolaides, Phys. Rev. A6, 2078 (1972).

94. "The Hartree-Fock Method for Atoms" C. Froese-Fischer
    (Wiley, New York, 1977).

95.   "Excited States in Quantum Chemistry" C. A. Nicolaides
      and D. R. Beck, eds. (Reidel, Dordrecht, Netherlands, 1978).

96.   "Particle-Hole Formulation of the Unitary Group Approach to
      the Many-Electron Correlation Problem" J. Paldus and M. J.
      Boyle, Phys. Rev. A22, 2299; 2316 (1980).

97.   "Specific Correlation Effects in Inner Electron Photoelectron
      Spectroscopy" D. R. Beck and C. A. Niclaides, Phys. Rev. A26,
      857 (1982).

98.   "Molecular Electronic Structure by Combination of Fragments"
      B. Kirtman, J. Phys. Chem. 86, 1059 (1982).

99.   "Principles for a Direct SCF Approach to LCAO-MO Ab Initio
      Calculations" J. Almlöf, K. Faegri and K. Korsell, J. Comp.
      Chem. 3, 385 (1982).

100. "K-Shell Binding Energy of Be and Its Fluorescence Yield"
      C. A. Nicolaides, Y. Komninos and D. R. Beck, Phys. Rev.
      A27, 3044 (1983).

101. "Density Dynamics: A Generalization of Hartree-Fock Theory"
      D. J. Rowe, M. Vassanji and G. Rosensteel, Phys. Rev. A28,
      1951 (1983).

102. "Density Matrix Treatment of Localized Electronic Interac-
      tions. Separated Electron Pairs" B. Kirtman, J. Chem. Phys.
      79, 835 (1983).

PARTICIPANTS

Reinhart Ahlrichs
Institut für Physikalische Chemie
Universität Karlsruhe
Kaiserstrasse 12
7500 Karlsruhe
FEDERAL REPUBLIC OF GERMANY

Jan Almlöf
Kjemisk Institut
Oslo Universität
Blindern P.O.B. 1033
Oslo 3
NORWAY

Raymond A. Bair
Chemistry Department
Argonne National Laboratory
Argonne, Illinois 60439

Rodney J. Bartlett
Department of Chemistry
University of Florida
Gainesville, Florida 32611

Charles W. Bauschlicher, Jr.
NASA Ames Research Center
Moffet Field, California 94035

J. Stephen Binkley
Theoretical Division
Sandia Laboratory
Livermore, California 94550

Thom. H. Dunning, Jr.
Chemistry Department
Argonne National Laboratory
Argonne, Illinois 60439

Clifford E. Dykstra
Department of Chemistry
University of Illinois
505 South Mathews Avenue
Urbana, Illinois 61801

Martyn F. Guest
SRC Daresbury Laboratory
Daresbury
Warrington WA4 4AD
UNITED KINGDOM

Dermot Hegarty
Chemische Laboratoria
der Rijksuniversiteit
Nijenborgh 15, 9747 AG
Groningen
NEDERLAND

Poul Jorgensen
Division of Theoretical Chemistry
Department of Chemistry
University of Aarhus
DK-8000 Aarhus C
DENMARK

Bernhard Levy
Laboratoire Chemie
L'Ecole Normal Superieur
1, Rue Maurice Anonx
F 92120 Mountronge
FRANCE

Bowen Liu
IBM Research Laboratory
San Jose, California 95193

Wilfried Meyer
Fachbereich Chemie
Universität Kaiserslautern
D-6750 Kaiserslautern
FEDERAL REPUBLIC OF GERMANY

Cleanthes A. Nicolaides
Theoretical and Physical
  Chemistry Institute
N.H.R.F.
48 Vas. Constantinou Avenue
Athens 501/1
GREECE

Josef Paldus
Department of Chemistry
University of Waterloo
Waterloo, Ontario N2L 3G1

Anthony K. Rappe
Department of Chemistry
Colorado State University
Fort Collins, Colorado 80523

V. R. Saunders
SRC Daresbury Laboratory
Daresbury
Warrington WA4 4AD
UNITED KINGDOM

Henry F. Schaefer III
Department of Chemistry
University of California
Berkeley, California 94720

Isaiah Shavitt
Department of Chemistry
Ohio State University
140 W. 18th Avenue
Columbus, Ohio 43210

Peter R. Taylor
CSIRO
Division of Chemical Physics
Clayton, Victoria
AUSTRALIA

H.-J. Werner
Los Alamos National Laboratory
Mail Stop J569
Los Alamos, New Mexico 97545

SPECIAL REPRESENTATIVES

John Gustafson
Floating Point Systems
P.O. Box 23489
Portland, Oregon 97223

T. A. Truelson
Science Administrator
NATO Scientific Affairs Division
B-1110 Bruxelles
BELGIUM

David Vickers
Floating Point Systems
P.O. Box 23489
Portland, Oregon 97223

INDEX